# Mushrooms and fungi

A Hamlyn Colour Guide

# Mushrooms and fungi

**By Dr Jaroslav Klán**

Illustrated by
Bohumil Vančura

**Hamlyn**
London • New York • Sydney • Toronto

Translated by Daniela Coxon
Designed and produced by Artia for
The Hamlyn Publishing Group Limited
London • New York • Sydney • Toronto
Astronaut House, Feltham, Middlesex, England



ISBN 0 600 352 88 9

Printed in Czechoslovakia
3/15/04/51-01

# Contents

## Introduction

Fungi (Mycota) form a large independent kingdom in the natural world alongside green plants and animals. Fungi were classified with the flowering plants, but their separate classification is justified by basic differences in the way they take in nourishment. In the first place fungi lack chlorophyll and therefore are unable to use the sun's energy. They are likewise unable to transform simple substances, such as carbon dioxide, water or minerals from the soil, into the organic matter of their bodies. Therefore they absorb food from organic matter which has been produced by autotrophic plants capable of photosynthetic assimilation. Fungi receive their nutrients from animal or plant bodies, either from dead organisms, as in the case of saprophytic fungi, or from living organisms, as in the case of parasitic fungi or mycorrhizal fungi living in symbiosis with plants. The physiology and morphology of fungi have become adapted to this way of feeding, so that all the necessary components, in particular carbonaceous substances, are received in the form of solutions from the soil. To break down and decompose vegetable or animal matter from the soil, fungi produce a large quantity of enzymes. In effect this means that the bodily function is completely antithetical to that of green plants. While green plants produce organic matter, fungi and other heterotrophic organisms decompose it. Minerals are thus returned to the soil and carbon dioxide to the atmosphere and in this way the continual circulation of matter and energy is maintained. Without this basic and irreplacable action of fungi, the Earth's surface would be littered with dead animals and vegetation and life itself would grind to a halt. There could be no life, therefore, without the existence of fungi. When a tree is felled, the stump is left in the ground. Within 50 years fungi will have transformed this hard, woody matter into humus. During this period of time the stump hosts dozens of arboreal species of fungi, which create a chain of activity that is constantly being repeated given similar conditions.

The total number of fungi is prodigious, reaching about 60, 000 species distributed over the entire planet. A large proportion of this number is accounted for by microscopic fungi. Large fungi, which can be easily identified by the human eye, are only represented by some 6,500 species, of which about 6,000 species are Basidiomycetes and the remainder are Ascomycetes.

The first fungi appeared on Earth probably more than a thousand million years ago. The earliest developmental stages of fungi and the first fossil finds date back to the late Palaeozoic. This was about

300 million years ago, when the first veined plants came into existence. The most abundant remnants of fungi have been preserved in coal deposits which are the fossil remains of carboniferous virgin forests. Coal contains the spores, filaments, tissue and fruit-bodies of a wide range of fungi of various groups, showing that fungi were abundant in the organic substrate of the carboniferous forests. They lived as parasites on the vegetation of that period and, together with bacteria, participated in its decomposition and transformation into coal.

## Large Fungi

The fungus-body (thallus) is either formed by a single microscopic cell, similar to those found in yeast fungi, or consists of many cells, which are either branched and filamentous as in moulds, or take the form of a variously shaped network of long, septate filaments (hyphae). The thallus of a large fungus is composed of long, thin, white, densely interwoven hyphae which penetrate the substrate. These filaments are known as the mycelium. Given favourable climatic and physiological conditions, the mycelium develops a reproductive organ, the fruit-body (carpophore). However, while the mycelium can exist in the same place for decades, as long as the environment undergoes no changes, fruit-bodies survive for a relatively short time. The fruit-bodies of some inky-caps *(Coprinus)* can develop within an hour and decompose soon after. The fleshy fruit-bodies of Boleti take ten to fifteen days to reach full size and their decay takes about the same time. Mushrooms with leathery or woody fruit-bodies, such as the Polyporaceae, live for a whole year or even longer, continuing to grow in favourable conditions. The decomposition of fruit-bodies is assisted by bacteria and moulds, but also by various insect larvae, slugs, squirrels and other animals. The fruit-bodies of some species, for example those of the Fairy-ring Champignon, shrivel during dry spells but swell to their original size after a period of rain.

## Reproduction of Fungi

The sole function of the fruit-body as a reproductive organ is to produce single-celled or multi-cellular reproductive formations, called spores. Spores are minute in size, measuring between 3 to 20 μm (thousandths of a millimetre). Larger sizes than this are excep-

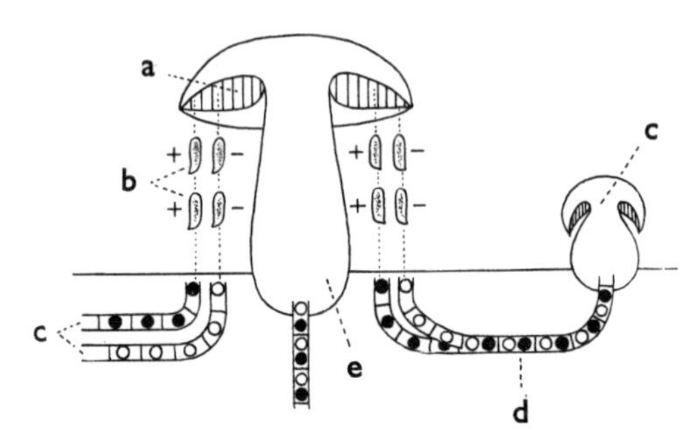


1. Reproduction of fungi: a — tubular hymenophore; b — genetic variations of spores; c — monokaryotic mycelium (primary mycelium); d — dikaryotic mycelium with clamp connections (secondary mycelium); e — flesh composed of dikaryotic hyphae

tional. Such measurements give the length of the spore in relation to its width. For example, in the specification 8—9.5/3—4.2 µm, the initial figures give the minimum and maximum length of the spore, while the figures following the oblique line refer to the minimum and maximum width of the spore. The shape, size and ornamentation of spores vary very little within a given species and are therefore of great importance in the classification of fungi.

Fruit-bodies release their spores only when ripe and then over a relatively short period of time. The number of spores produced by a single fruit-body is enormous. In the case of the Common Many-zoned Polypore *(Trametes versicolor),* a surface of 10 cm² ejects some 1.4 thousand million spores over a period of 24 hours. The 1 cm-wide fruit-body of a Gill fungus (genus *Crepidotus* ) discharges 1.3 thousand million spores in the same period of time. Although individual spores are visible only under a microscope, the spore powder (deposit), which has a characteristic colour, can be seen by the naked eye. Because of their minute size, spores are easily transported by wind over large distances. However, the same fungus species do not necessarily grow on all continents. In fact such cosmopolitan species as *Mycena pura,* for instance, are rare.

The germination and growth of a spore into a single filament, known as a uninucleate or primary mycelium, is only possible under certain constant conditions, which are determined by humidity, temperature and the character of the soil. Most fungus species require two primary mycelia with genetically distinct nuclei to merge and form a two-nucleus mycelium (secondary mycelium). Such mycelium can be found if the surface layer of humus around a fruit-body is removed. In certain conditions the mycelium produces nodules, primordia, which are the basis of future fruit-bodies. The closely interwoven and crowded filaments of the mycelium can develop into hard, root-like structures called rhizomorphs. These can be variously shaped; for example, they are mesh-like in the Honey Fungus, bristlelike in *Marasmius scorodonius*, or forming hard, globular sclerotia in *Collybia tuberosa.*

In Ascomycetes the thecium is usually found on the surface of the fruit-body, for example in *Gyromitra esculenta.* Spores are formed inside the ascus, usually eight at a time. They are then released when the ascus opens through a slit or a lid. The space between the asci is filled by the sterile tips of hyphae, known as paraphyses.

In Basidiomycetes the hymenium is either on the surface of the gills (in the Gill fungi, for instance *Amanita* species) or in the tubes (in Boleti and Polyporaceae). In other cases it can cover almost all the smooth surface of variously branched fruit-bodies (as in *Ramaria* ), envelop the surface of the spikes (as in *Hydnum* ), or be enclosed by globular fruit-bodies (as in *Lycoperdon* ). The spores are formed, usually in fours, on the surface of club-shaped cells called basidia. Before they have ripened they are usually connected with the basidia by a stalk called the sterigma. When mature they are then ejected into the air.

Apart from spores which have been formed sexually, there are some spores whose origin is asexual. This phenomenon is very rare in large fungi, but is quite common of microscopic fungi. Fungi can also reproduce vegetatively from segments of a mycelium.

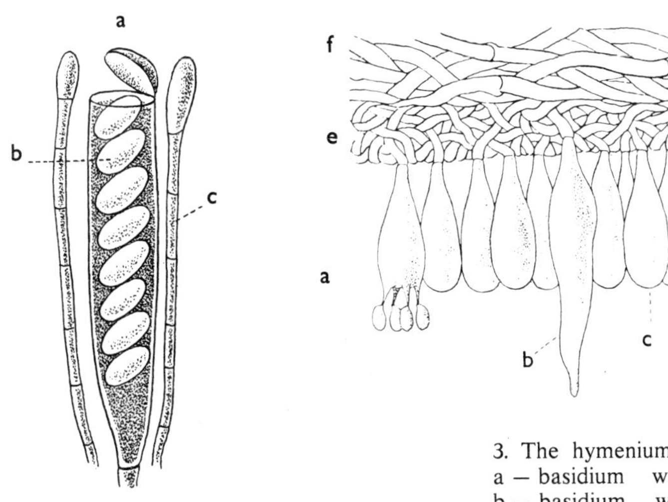


2. Ascus (a) with spores (b) and paraphyses (c)

3. The hymenium of large fungi a — basidium with 4 spores; b — basidium with sterigmata (stalks); c — basidioles; d — cystidia; e — subhymenium; f — hyphae with clamp connections

## Morphology and Anatomy of Fungi Fruit-bodies

In large fungi, the shapes of fruit-bodies vary greatly. The variation is largely influenced by their age, weather conditions, environmental and other factors. The size of fruit-bodies usually falls within certain limits, but in some cases the average size can be exceeded.

The shape of the fruit-body can be either simple or complex; it can be bowl-shaped, cup-shaped, chalice or ear-like (as in *Aleuria, Otidea*), or club-shaped and branching *(Ramaria, Sparassis),* cornet-like *(Craterellus),* star-shaped *(Geastrum),* globular or pear-shaped *(Bovista, Lycoperdon),* hoof-shaped *(Piptoporus),* fan-shaped *(Pleurotus),* or can be divided into an upper section formed by the cap and a lower section consisting of the stipe *(Amanita, Boletus).*

The size of the cap varies from one millimetre in some *Marasmius* species to several hundreds of millimetres in the Parasol Mushroom *(Macrolepiota procera).* They can be lobed *(Gyromitra)* or ribbed *(Morchella),* or with a broad expanse *(Pluteus),* semi-globular *(Boletus),* sharply or bluntly conical *(Hygrocybe),* bell-shaped or helmet-shaped *(Mycena),* funnel-shaped *(Clitocybe),* cylindrical or ovate *(Coprinus).* The margin is straight, wavy, sharp, blunt, striate or curled. The surface of the cap has also particular characteristics, being either smooth, wrinkled, striate, felt-like, fibrilous or scaly, slimy, sticky, glossy or dull. The caps of some species absorb water when damp and turn darker in colour. In the process of drying out, they again become lighter, starting from the centre. Such caps are called hygrophanous *(Kuehneromyces mutabilis* or some *Clitocybe* species).

The underside of the cap is usually covered with spikes, tubes or gills, which are coated by the hymenium. Spikes are the least common; they vary in length, width, thickness, and density. Tubes have probably developed from gills, which became connected by transverse partitions. Openings of tube-pores are either elongated, irregu-

4. Various shapes of fruit-bodies: a, b, c — fruit-bodies consisting of cap and stipe; d — cornet-shaped; e — branching; f — hoof-shaped; g — lobed; h — pear-shaped

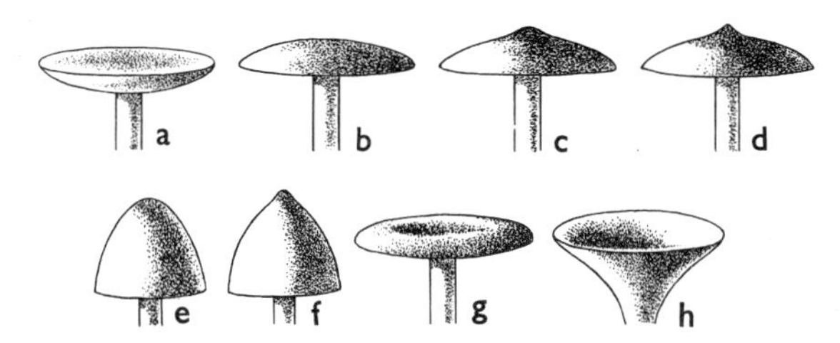


5. Various shapes of caps: a — expanded; b — convex; c — expanded with umbo; d — expanded with papilla; e — bell-shaped; f — club-shaped; g — expanded with central depression; h — funnel-shaped

larly angular or even regularly rounded. They vary in height, width and the thickness of their walls. They are attached to the stipe in a number of ways; they can be remote *(Leccinum)* or adnate *(Xerocomus)* or decurrent *(Boletinus).* The shorter tubes resemble holes in a net and form a delicate, variously coloured lacework on the stipe.

Gills or lamellae are flat, radiating scales, which are directed from the cap's margin towards the stipe. Long gills reaching the stipe are often interspersed with shorter gills. Gills are either simple or forked; sometimes they are connected by transverse ribs. One of the most characteristic features of some genera is the way in which the gills are attached to the stipe. Gills vary in height and thickness. For example, the *Pluteus* genus has very thin yet dense gills, while those of *Russula nigricans* are thick but broadly spaced. The colour of the gills is determined by that of the fruit-body flesh and by the colour of the spores on the gill's surface. For example the gills of some *Agaricus* species are white or whitish when young, because the spores have not yet ripened. The spore powder is black-brown when ripe and then the gills acquire a similar colouring. However, the purple or fleshy colouring of the gills in the *Laccaria* genus is determined by the colour of the flesh, as the spore powder is white. Thus the colour of the gills does not always reveal the colour of the spore powder. This can be determined by placing the fruit-body on a sheet of white paper and covering it. After a few hours the ripe spores will fall out and create an exact, evenly coloured pattern of the gills or spores. The spore powder print is an important characteristic feature for the systematic classification of some groups of fungi. It can be white, yellow, pink, brown or black-brown. The edges of the gills are either smooth, serrate or undulate, and in some species their colour differs from the colour of the face of the gills. In *Pluteus atromarginatus* the gills are pink but their blades are black. Another important feature of the gills is their texture; some are fragile and break easily *(Russula),* others are waxy *(Hygrocybe),* or elastic *(Marasmius),* or are pliable, as if greasy *(Russula cyanoxantha);* some gills liquify when they ripen *(Coprinus).* In some genera only low ridges or veins are formed instead of gills.

The stipe's position in relation to the cap is either a central, eccentric or lateral one. In shape, it varies a great deal. It can be long and thin like a hair, as in the case of some *Marasmius* species, or several centimetres thick as it is in some Boleti. Its surface structure very frequently resembles the surface of the cap. The size of the stipe is given in two measurements; the first is the height (length) and the second is the width (thickness). For example, there might be a length of 5—10 cm and thickness of 1—2 cm.

The flesh of the fruit-body is a basic feature which determines the durability and the quality of the mushroom. The nature of the flesh is determined by its anatomical structure, but it undergoes important changes in the course of the development of the fruit-body. For example young fruit-bodies of the Brown Birch Boletus *(Leccinum scabrum)* have a delicate structure and pleasantly aromatic flavour, while adult specimens have woody and indigestible stipes and mushy caps saturated with water. There are five basic types of mushroom flesh, namely, gelatinous *(Tremella),* juicy and fleshy *(Boletus),* fragile *(Russula),* cottony *(Agaricus),* or leathery to woody (the majority of the Polyporaceae). Apart from many exceptions, there are also many transitional categories between the above types.

The cap and stipe of most mushrooms bear vestigial traces and remnants of other parts of the fruit-body, which characterize its early growth, namely the veil (velum). We can distinguish the universal veil (velum generale) and the partial veil (velum partiale). The veil is particularly noticeable in the *Amanita* species. When young this mushroom is ovate and completely covered by the universal veil. As it grows, the veil breaks and the adult fruit-body displays its remnants only around the base of the stipe and in the form of warts, on the surface of the cap. The gills are covered by a partial membranous veil which in adult fruit-bodies remains in the shape of a ring around the stipe and in patches along the margin of the cap. In some species the veil is cottony, cobweb-like or slimy. These veil remnants are prominent in *Agaricus, Lepiota* and some *Suillus* species.

The hyphae in a fruit-body are arranged sequentially, forming cords or variously interwoven filaments. The majority of soft, fleshy fruit-bodies consist of filaments of a single type, but fruit-bodies with a tough, woody consistency have two or three different types of filaments. The filaments are often equipped with clamp-like connections which stretch from one cell to the next. They originate by cell division. The hymenium always contains basidia with stalks and spores and infertile basidia without stalks. Special cells can also be present which grow from the subhymenium and protrude above the basidia.

These are called cystidia and their variable shapes are often important criteria in classification. Cystidia which are located on the gill edges are called cheilocystidia; those found on the flat surface of gills are known as pleurocystidia. Cystidia are also visible on the surface of the cap and the stipe. They can be empty, filled or encrusted.

Sexually produced spores have a different shape, size, colouring and surface structure. A spore cell is enveloped by a wall which consists of several layers. The underside of a spore bears a visible protrusion (apiculus), which originally attached the spore to the basidium stalk. In some genera (e.g. *Coprinus, Stropharia* and *Pholiota*) the opposite end is depressed to form a germinating spore. This is the place where the young hypha grows outwards and where the spore germinates.

## Systematic Classification of Fungi

All fungi are classified on the basis of morphological, anatomical, physiological and biochemical features into a particular system. Individuals which have certain morphological features in common form a species. Every species has its own area of distribution and differs from other species in its ecology and genetic characteristics. A large grouping, which encompasses the species, is the genus. Genera are then grouped into families, followed by orders and classes.

The following brief systematic survey outlines the characteristics of the families and some genera of the fungi which are described and illustrated in this book.

## Key to the Classification of the Main Families

A. Fleshy fruit-body composed of a cap and a stipe. Hymenium on the surface of true or reduced gills.

1. Spore powder white, whitish, yellow or ochre: *Pleurotaceae, Hygrophoraceae, Tricholomataceae, Amanitaceae, Agaricaceae, Paxillaceae, Cantharellaceae, Russulaceae* families.

   *Pleurotaceae:* whitish gills, decurrent to the stipe; stipe usually lateral or eccentric.
   *Hygrophoraceae:* fleshy, soft, juicy or even watery fruit-bodies; gills thick, sparse and waxy.
   *Tricholomataceae:* a diverse family, difficult to characterize. Gills are never free but attached in various ways to the stipe.
   *Amanitaceae:* free gills, veil is often well developed.
   *Agaricaceae:* a diverse family; gills free; stipe easily detached from cap; veil is developed.

*Paxillaceae:* fleshy fruit-bodies and decurrent gills, yellow-brown or orange in colour.
*Cantharellaceae:* pileate or funnel-shaped fruit-bodies; hymenophore composed of ridges, veins or wrinkles.
*Russulaceae:* fruit-bodies with fragile flesh, distinguished by globular cells, gills adnate.

2. Spore powder pink or pinkish: *Pluteaceae, Entolomataceae* families; genus *Lepista* and some representatives of the genus *Laccaria.*
*Pluteaceae:* fleshy fruit-bodies and free, fleshy-pink gills.
*Entolomataceae:* pink gills, attached in a variety of ways but never free.
3. Spore powder black to olive green: *Coprinaceae, Gomphidiaceae* families, and genus *Agaricus.*
*Coprinaceae:* fragile, fruit-bodies with brief developmental phase; gills free or attached.
*Gomphidiaceae:* fleshy fruit-bodies with thick, decurrent gills.
4. Spore powder rusty brown to brown: family *Cortinariaceae,* genus *Paxillus* and exceptionally genus *Hypholoma.*
*Cortinariaceae:* this family cannot be simply classified as it is heterogeneous. Fruit-bodies of the majority of species have a well developed veil.
5. Spore powder is dark purple to purple-black; veil either present or absent: *Strophariaceae* family.

B. Fleshy, pulpy fruit-body, consisting of a cap and stipe; hymenium located inside the tubes; spore powder almost olive green: *Boletaceae* family.
C. Fruit-body cork-like, hard, sometimes fragile, usually hoof or fan-shaped; cap and stipe rare; hymenium inside tubes; spore powder white to yellowish: *Polyporaceae* family.
D. Fruit-body pulpy, fragile, sometimes divided into a cap and a stipe; hymenium covers surface of spikes; spore powder white or yellowish: *Hydnaceae* family.
E. Fruit-body laminated, gelatinous in damp conditions, irregularly lobed, elastic, not divided into cap and stipe; hymenium on the surface; spore powder white: *Auriculariaceae, Tremellaceae* families.
F. Fruit-body fragile, undivided into cap and stipe, simple club-shaped or branched like a bush; hymenium on the surface; spore powder whitish or yellowish: *Clavariaceae, Ramariaceae* families.
G. Fruit-body fragile, bowl or ear-shaped, adnate or hollow, divided into stipe and undulate, globular cap; hymenium on surface; spore powder is white to ochre: *Otideaceae, Helvellaceae, Morchellaceae* families.
H. Fruit-body undivided into cap and stipe; hymenium enclosed by casing: *Phallaceae, Sclerodermataceae, Lycoperdaceae, Geastraceae* families.
*Phallaceae:* cylindrical or globular fruit-bodies with a gelatinous casing when young. When casing splits, it releases a porous, hollow formation with spores at apex in a foul-smelling, slimy layer.
*Sclerodermataceae:* globular fruit-bodies, black spore powder. Ripe fruit-bodies burst open with an irregular shape.
*Lycoperdaceae:* globular or pear-shaped fruit-bodies; external casing soon becomes detached but remains as an internal, paper-like peridium. Spore powder dark brown.
*Geastraceae:* globular fruit-bodies at first; external leathery casing bursts into star-shaped corners, each carrying internal fertile layer (gleba). Spore powder is brown.

# COLOUR PLATES

## Common Morel — Morels

*Morchella esculenta* L. ex ST. AM. — *Morchellaceae*

The Morels, together with the Brain-fungi, *Verpa* species and False Morels, belong to the Cup fungi. The Morels form fruit-bodies, mainly in April and May. They vary greatly in terms of colour and shape and in the past scientists were sidetracked into naming individually every single shape and colour variety. The original estimate of 30 species has been reduced to 6 to 8 worldwide.

The Common Morel is the most widespread representative. It has hollow, fragile fruit-bodies; the cap is globular, with irregularly arranged pits and ribs; it is light brown to black; the stipe is sometimes rugose. As a saprophytic mushroom the Common Morel can be found in all habitats, with sufficient organic matter in the soil, e. g. in well-lit deciduous forests, along their edges, in gardens and parks, in thickets and along the edges of wasteland. It occurs in large numbers from the lowlands to the highlands of the whole temperate zone of the northern hemisphere, from Japan as far as North America and also in Australia.

Related species, such as *Morchella conica* and *M. elata,* have a different shape and colour, and regularly ribbed caps. They grow in similar habitats to the Common Morel. The Morels can be confused with the poisonous *Gyromitra esculenta.* However *Gyromitra* caps have brain-like lobes instead of ribs. All Morels are edible and tasty.

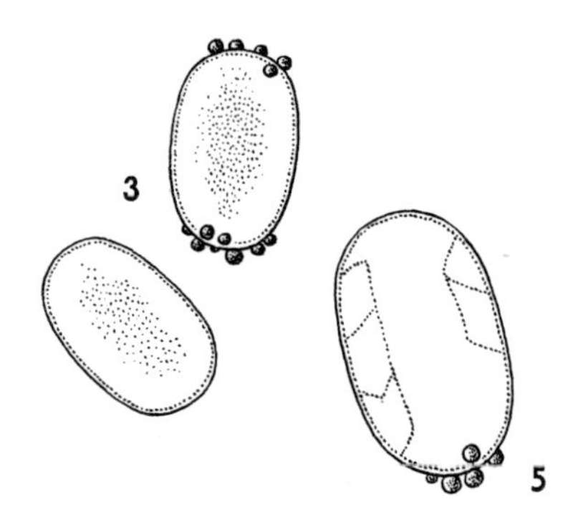


*Morchella esculenta* (1) has cap 3—5 cm high, 3—6 cm wide, ochre to light brown. Stipe 5—9 cm long, 2—3 cm thick, whitish. Flesh whitish (2). Taste and smell mild. Spore powder is light ochre. Cylindrical asci open through a small lid, containing 8 colourless, smooth spores (3), 18—20/10—13 µm, whithout oily droplets.

*Morchella conica* (4) has cap 4–6 (10) cm high, 3–5 (8) cm wide, rusty grey, blackish. Stipe 3–5 cm long, 1.2–2 cm thick, whitish. Flesh whitish. Taste and smell mild. Spore powder ochre; shape of asci and spores closely resembles that of the Common Morel; spores 20–32/13–16 μm in size (5).

## Brain Mushroom

*Helvellaceae*

*Gyromitra esculenta* (PERS. ex PERS.) FR.

The Brain Mushroom *(Gyromitra)* and the False morels *(Helvella)* belong to the same family. There are about eight *Gyromitra* species worldwide, half of which grow in Europe. *Gyromitra esculenta* is the most abundant. Its gregarious fruit-bodies grow mainly in spring, at the latest in April. The cap is sub-globose, with irregular brain-like lobes on the surface and hollow inside. The stipe is often somewhat grooved. *G. esculenta* grows as a saprophyte in mountain spruce forests and plantations and in pine forests on a sandy base, most frequently in bare soil. It is widespread throughout the temperate zone of the northern hemisphere. *G. esculenta* contains protein substances, gyromitrin and methylhydrazin, which can cause poisoning. It is possible to collect fruit-bodies which are several weeks old, and these overripe fruit-bodies have a higher poison content which can be fatal if eaten. The poison partly decomposes with thorough boiling; nevertheless this mushroom cannot be recommended for eating.

The closely related and very similar *Neogyromitra gigas* also grows in spring, but in deciduous forests around tree stumps. Externally it differs in the lighter colour of its caps, microscopically in the size and shape of its spores. *Gyromitra infula* and *G. fastigiata* are rare species, which grow in autumn in damp forests. All three species are edible and have a good flavour.

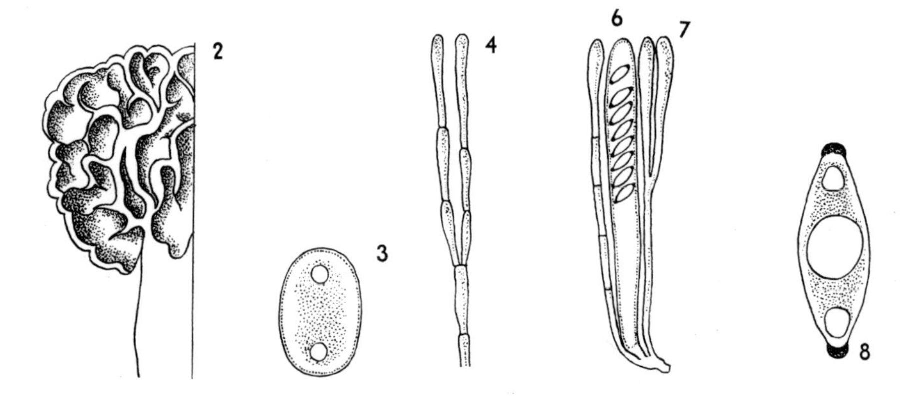

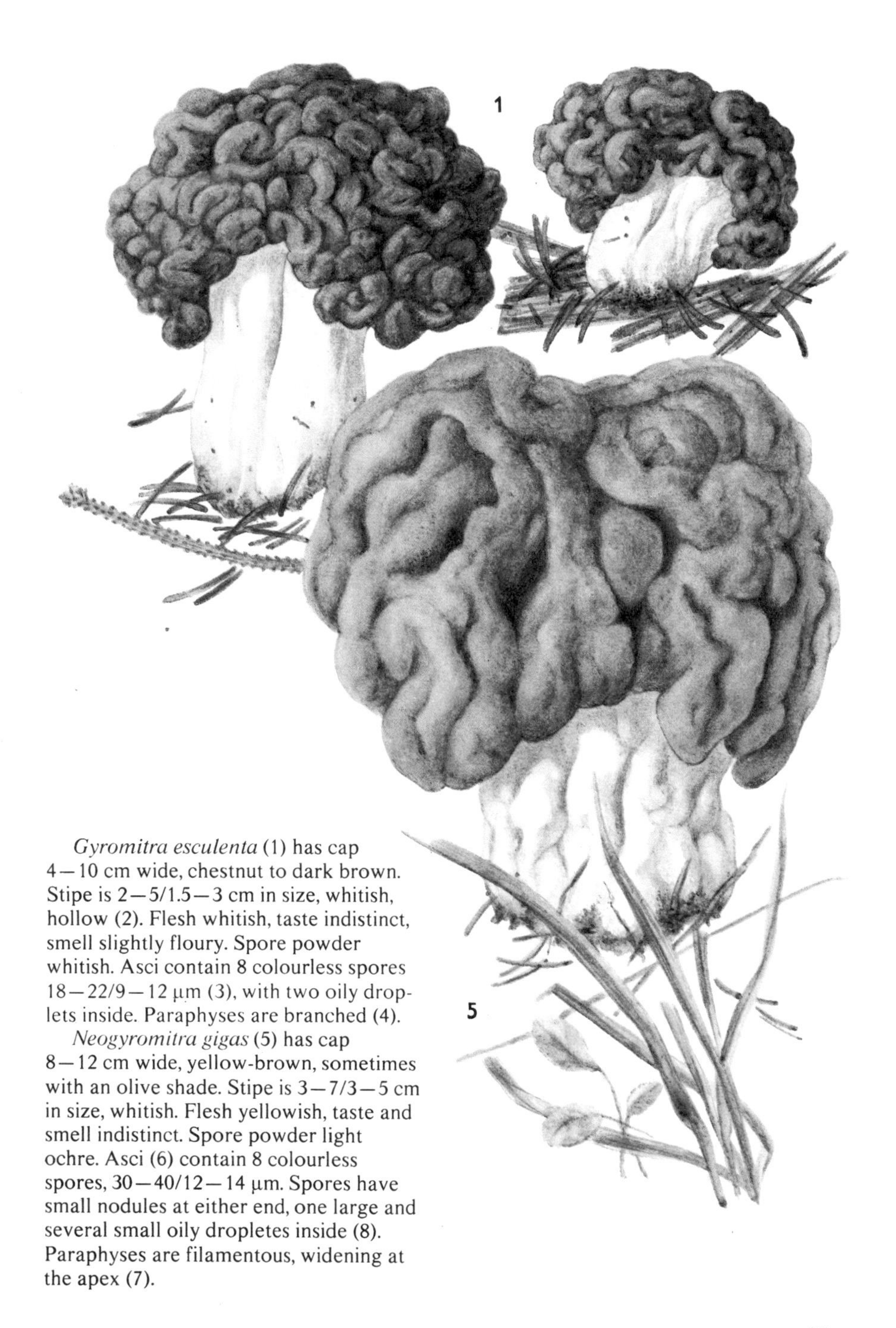


*Gyromitra esculenta* (1) has cap 4—10 cm wide, chestnut to dark brown. Stipe is 2—5/1.5—3 cm in size, whitish, hollow (2). Flesh whitish, taste indistinct, smell slightly floury. Spore powder whitish. Asci contain 8 colourless spores 18—22/9—12 μm (3), with two oily droplets inside. Paraphyses are branched (4).

*Neogyromitra gigas* (5) has cap 8—12 cm wide, yellow-brown, sometimes with an olive shade. Stipe is 3—7/3—5 cm in size, whitish. Flesh yellowish, taste and smell indistinct. Spore powder light ochre. Asci (6) contain 8 colourless spores, 30—40/12—14 μm. Spores have small nodules at either end, one large and several small oily dropletes inside (8). Paraphyses are filamentous, widening at the apex (7).

## Orange-peel Fungus



*Aleuria aurantia* (PERS. ex HOOK.) FUCKEL

Both the Orange-peel Fungus and the Lemon-peel Fungus *(Otidea onotica)* grow abundantly in the entire temperate zone of the northern hemisphere, extending from Japan to the United States of America. They belong to the Cup fungi and both are soil saprophytes. The Orange-peel Fungus prefers sandy soil in damp meadows, gardens, orchards and parks and also grows in well-lit forests, particularly in beech woods and along their warm, sunny edges. The Lemon-peel fungus is most abundant in coniferous forests, but it can also be found in woods. Both species grow throughout summer and autumn; the Orange-peel Fungus often grows in spring, when its striking deep orange fruit-bodies appear in small clusters, particularly after rainfall. Both species are edible.

The genus *Aleuria* is not a large one; only four of its species grow in Europe. *Sarcoscypha coccinea* closely resembles the Orange-peel Fungus, but is has a scarlet thecium (hymenium), and its small, stalked fruit-bodies grow early in spring on the branches of deciduous trees. The genus *Otidea* consists of a greater number of species; about 12 can be found in Europe. Some can be relatively easily distinguished by the colour of their thecium; *O. leporina* has a rusty brown thecium, *O. abietina* brown with a purple tint, *O. umbrina* chocolate brown, *O. alutacea* dirty ochre and *O. concinna* beige yellow. They are edible.

3

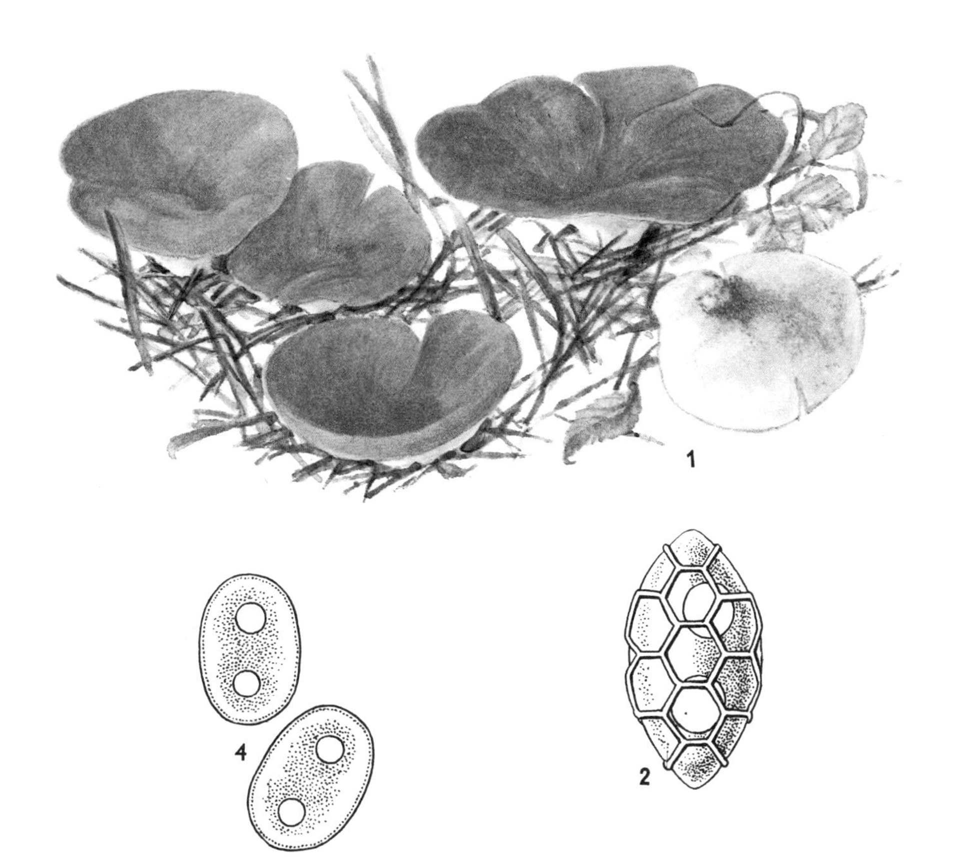


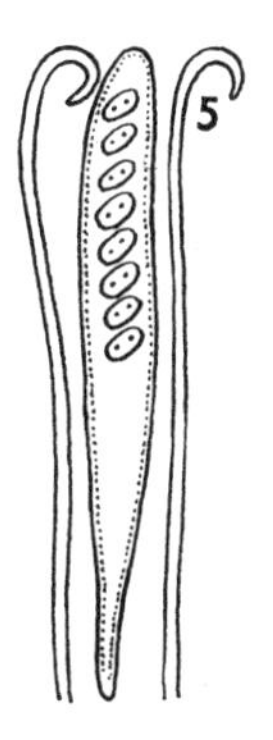


*Aleuria aurantia* (1) has cup-shaped fruit-bodies (apothecium) 2—10 cm wide, later expanded and wavy. Thecium orange or carmine red, whitish and downy outside; flesh grey-white and fragile. Taste and smell indistinct. Spore powder is white. Spores (2) are 17—24/9—11 μm, colourless, reticulate on the surface, with two droplets of oil.

*Otidea onotica* (3) has fruit-bodies 3—8 cm high, 1—3 cm wide, ear-shaped with inrolled margin. Thecium yellow-orange, yellow or ochre and smooth outside; flesh is waxy fragile. Taste and smell pleasant. Spore powder white; spores (4) are 10—13/5—6 μm, colourless, smooth, with two oily dropletes. Paraphyses are hooked (5).

## Jew's ear Fungus

*Auriculariaceae*

*Hirneola auricula-judae* (BULL. ex ST. AM.) BERK.

With the exception of some Scandinavian countries, the Jew's ear Fungus is almost cosmopolitan. It is saprophytic but can also be parasitic on the branches and stems of the elder, false acacia, maple, alder and beech. In North America it is most abundant on coniferous trees. The gelatinous and flexible fruit-bodies are ear or cup-shaped and undulate, and laterally attached to the wood. During drought they are hard and fragile but in damp conditions regain their original shape and constistency. They grow abundantly in clusters in autumn and mild winters. The Jew's ear is an edible mushroom, very popular in China and Japan where it is used in salads and as a vegetable.

In wet weather, a newly developed fruit-body of *Tremella foliacea* has numerous brain-like convolutions. In dry weather the lobes are small and hard but partly revive when moistened. *T. foliacea* grows on dead wood of deciduous trees, on beech and oak in the entire temperate zone of the northern hemisphere. *T. mesenterica,* with bright golden yellow fruit-bodies, is also quite abundant. Some species of this genus are parasites on other fungi *(T. encephala, T. mycophaga).* They are unsuitable for consumption.

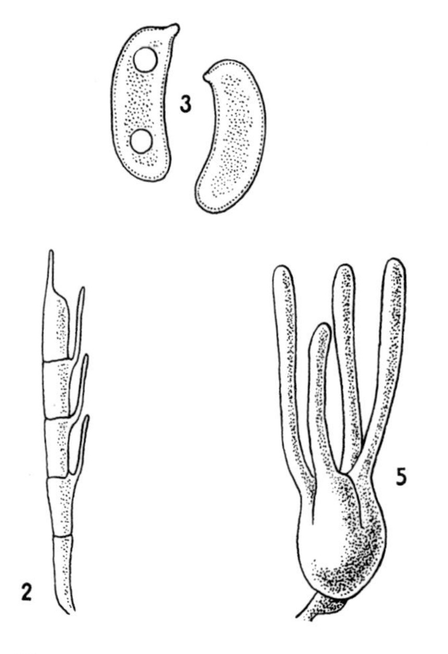


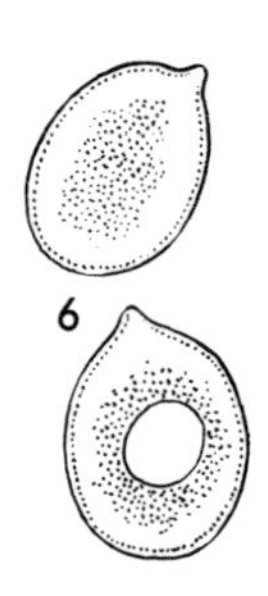


*Hirneola auricula-judae* (1) has fruit-bodies 3—10 cm wide, light brown or brown-yellow. Upper surface grey-brown and velvety; hymenium glossy brown, reticulate and smooth. Basidia have transverse divisions (2). Spore powder whitish; spores (3) are colourless, 13—23/4—8 µm.

*Tremella foliacea* (4) has fruit-bodies 3—20 cm wide, yellow-brown or brown; they are composed of convoluted, slightly translucent lobes, which are hard and small in drought. Taste and smell of both species are nondescript. Hymenium covers the entire surface of the lobes. Basidia have longitudinal divisions (5). Spore powder whitish; spores (6) colourless, 7—13/5.5—9 μm.

## Beautiful Clavaria

*Ramaria formosa* (PERS. ex FR.) QUÉL.

Fairy Clubs

*Ramariaceae*

The Fairy Clubs have densely branched and brightly coloured fruit-bodies (yellow, ochre, red, purple). They grow as saprophytes on forest humus or completely rotten wood. There are about 40 species worldwide, some of which are poisonous. *Ramaria formosa* and *R. aurea* are among the most abundant species in the temperate zone of the northern hemisphere.

*R. formosa* grows gregariously in the autumn in deciduous, particularly oak, woods, less frequently in coniferous forests, stretching from lowlands to highlands. Characteristic of this species are the lemon yellow tips of branches, which become paler with aging, and white flesh, which turns rusty or black when bruised or cut. The bitter taste of the fruit-bodies is strongest at the ends of their branches. *R. formosa* is considered slightly poisonous, because it sometimes causes stomach upset. *R. ignicolor* is similar in colour but half in size with very slender branches.

The Golden Coral Fungus *(R. aurea)* grows in summer and autumn, scattered in deciduous or less frequently in coniferous forests from lowlands to highlands. Fruit-bodies are entirely lemon yellow or yellow-ochre, and edible although their flesh is tough. The closely related *R. flava* has paler, sulphur yellow fruit-bodies, with a whitish stipe which turns brown-red when bruised; it is also edible.

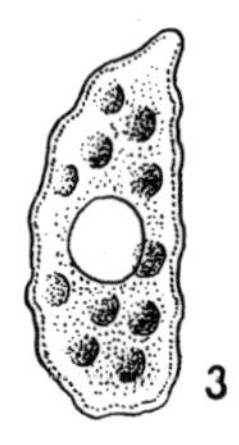


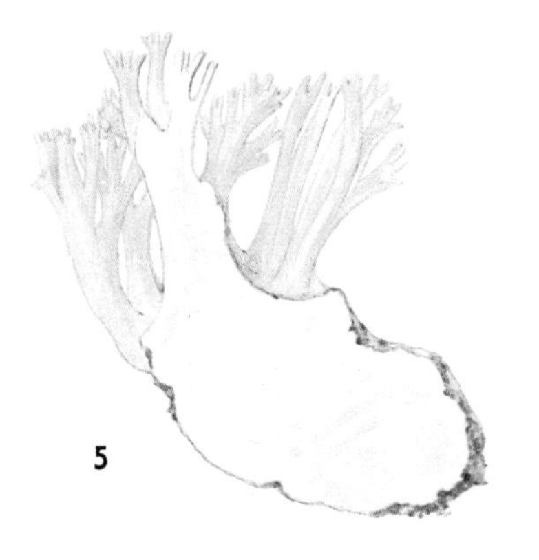

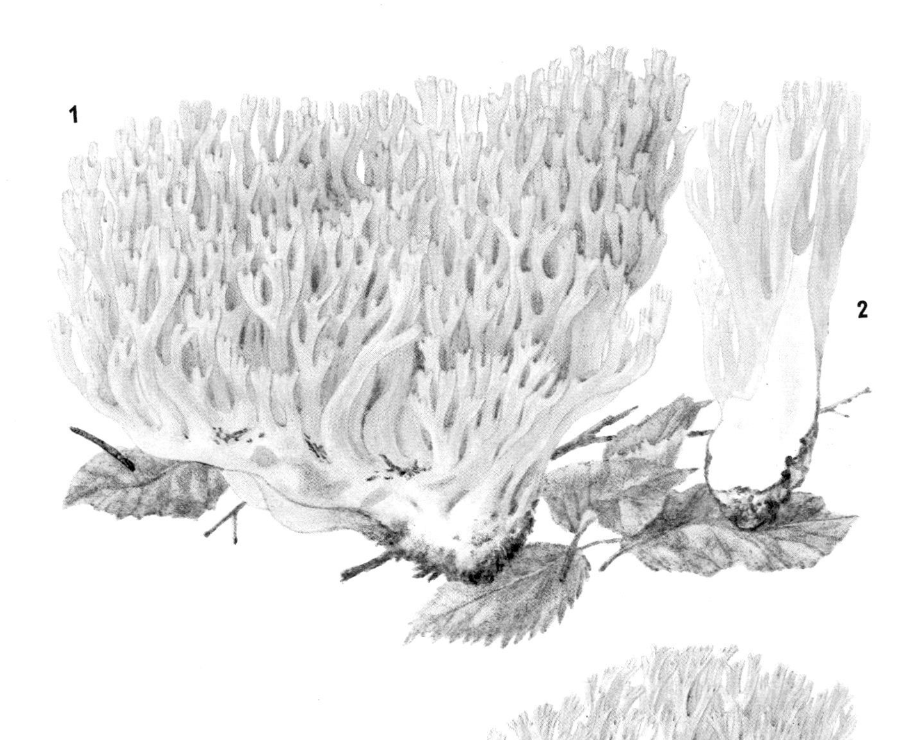


4

*Ramaria formosa* (1) has fruit-bodies 6—15 cm wide, 8—25 cm tall, composed of grooved, denticulate branches projecting from one stout stipe. Fruit-bodies yellow-red or yellow-ochre; teeth lemon yellow at the tips; flesh (2) white, rusty or black later; taste bitterish, aroma indistinct. Spore powder red-brown; spores (3) yellow, 8.5—15/4—6 μm.

*R. aurea* (4) has fruit-bodies 5—14 cm wide, 5—15 cm tall, composed of smooth denticulate branches, which grow out of a thickened stipe. Flesh (5) whitish, taste indistinct, aroma slightly floury. Spore powder yellow-ochre; spores (3) yellow, 11—13/4.5—6 μm.

## Crested Clavaria

Fairy Clubs

*Clavulina cristata* (HOLM. ex FR.) SCHROET.

*Clavariaceae*

The Crested Clavaria is the most common species of the genus *Clavulina.* It is a saprophytic species, most abundant in coniferous forests, but growing also in deciduous woods and in meadows and pastureland. It is widespread in the entire temperate zone of the northern hemisphere. This species is well distinguished by the sharp points at the ends of its branches, but its shape and colouring varies a great deal. Its bitterish taste makes it inedible. The related species *C. cinerea* with grey or smoky, purple tinged fruit-bodies, and *C. amethystina* with purple or lilac colouring, have small branches with blunt ends. Both species are edible.

The genus *Clavulinopsis* closely resembles the genus *Clavulina* but has a greater variety of species; almost 30 can be found in Europe. *C. corniculata* grows abundantly throughout the temperate zone of the northern hemisphere in meadows, pastureland and along grassy tracks. It is inedible due to its bitter taste. The denticulate branch tips characteristic of *C. corniculata* are absent in other related species. *C. umbrinella* has white, later pale brown or umber fruit-bodies and *C. luteo-ochracea* has cream or yellowish, almost unbranched fruit-bodies.

*Clavulina cristata* (1) has clustered or single fruit-bodies, 3—8 cm high, fragile; the branching tips are compressed and carry small combs with sharp ends (2). White, whitish or pinkish in colour. Flesh white; taste mild, later bitterish; aroma indistinct. Spore powder white; spores (3) are 8—10 µm, colourless with one large droplet of oil. Basidia have two stalks (4).

*Clavulinopsis corniculata* (5) has clustered or single fruit-bodies, 3—8 cm high, 2—3 times divided into elongated, finger-like branches (6). Fruit-bodies are slimy when damp, yolk yellow or ochre yellow in colour. Flesh ochreous (7); taste acrid or bitter, aroma floury. Spore powder white; spores (8) are 4—6 µm, colourless with one large droplet of oil. Basidia have four stalks (9).

1
2
4
3
9
8
5
6

## **Crisped Sparassis** *Sparassidaceae*

*Sparassis crispa* (WULF. ex FR.) FR.

*Sparassis crispa* is a parasitic mushroom widespread in the entire temperate zone of the northern hemisphere. Its fruit-bodies can be found in summer around the base of coniferous trees, most frequently pines and sometimes spruces and larches. The mushroom causes serious decay of roots and wood. *S. crispa* is edible, with a very pleasant flavour. Its friut-bodies, often weighing 2—3 kg, can be stored in cool places for some time without being attacked by moulds, as the mushroom contains sparassol, a substance with antibiotic properties. Fruit-bodies stored in a damp place seem to grow because their wavy, crisped branches stretch out and become broader. The related species *S. laminosa* is parasitic on oak trees; it is rare and occurs only in Europe. It can be recognized by its straight, flat and very broad branches.

The Bear's Head Fungus *(Hericium clathroides)* is a profusely branched mushroom covered with spikes. It grows in summer and autumn as a parasite and saprophyte on the wood of deciduous trees. In Europe it infests beeches, ashes, oaks and elms and in North America various species of poplars. In Europe the dead stumps of fir trees host a related species, *H. coralloides,* which has clusters of 1—2 cm long spikes at the tips of its branches. In North America this species grows on various deciduous trees. All *Hericium* species are edible.

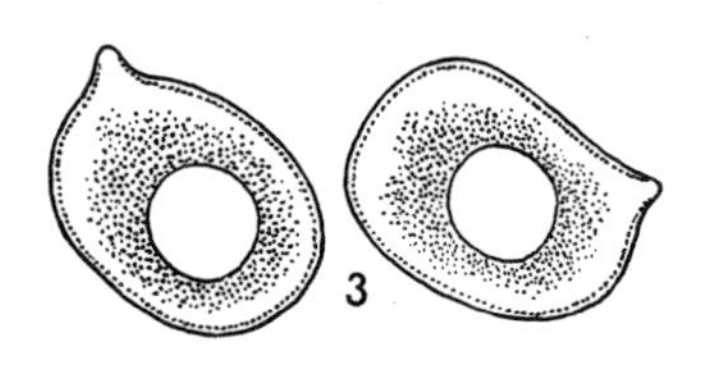


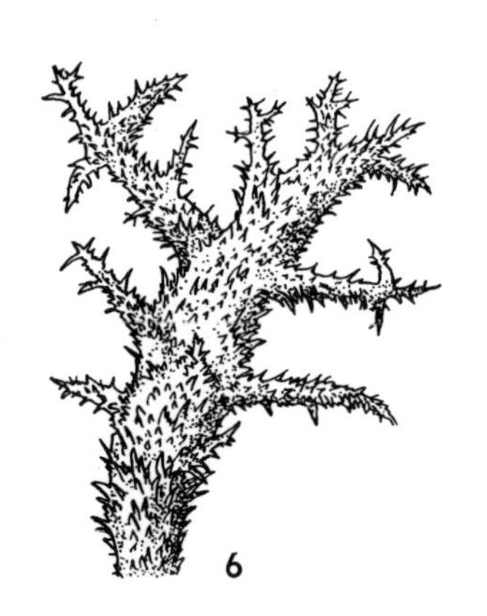


*Sparassis crispa* (1) has fruit-bodies 5—26 cm high, 6—30 cm wide, subglobose, divided into flattened, wavy lobes (2) which bear hymenium underneath. The white stipe is hidden in the ground. Fruit-bodies creamy, whitish or ochre. Flesh white; taste and smell pleasant, slightly spicy. Spore powder ochre; spores (3) 6—7/4—5 μm, pale yellow or colourless and have an oily droplet.

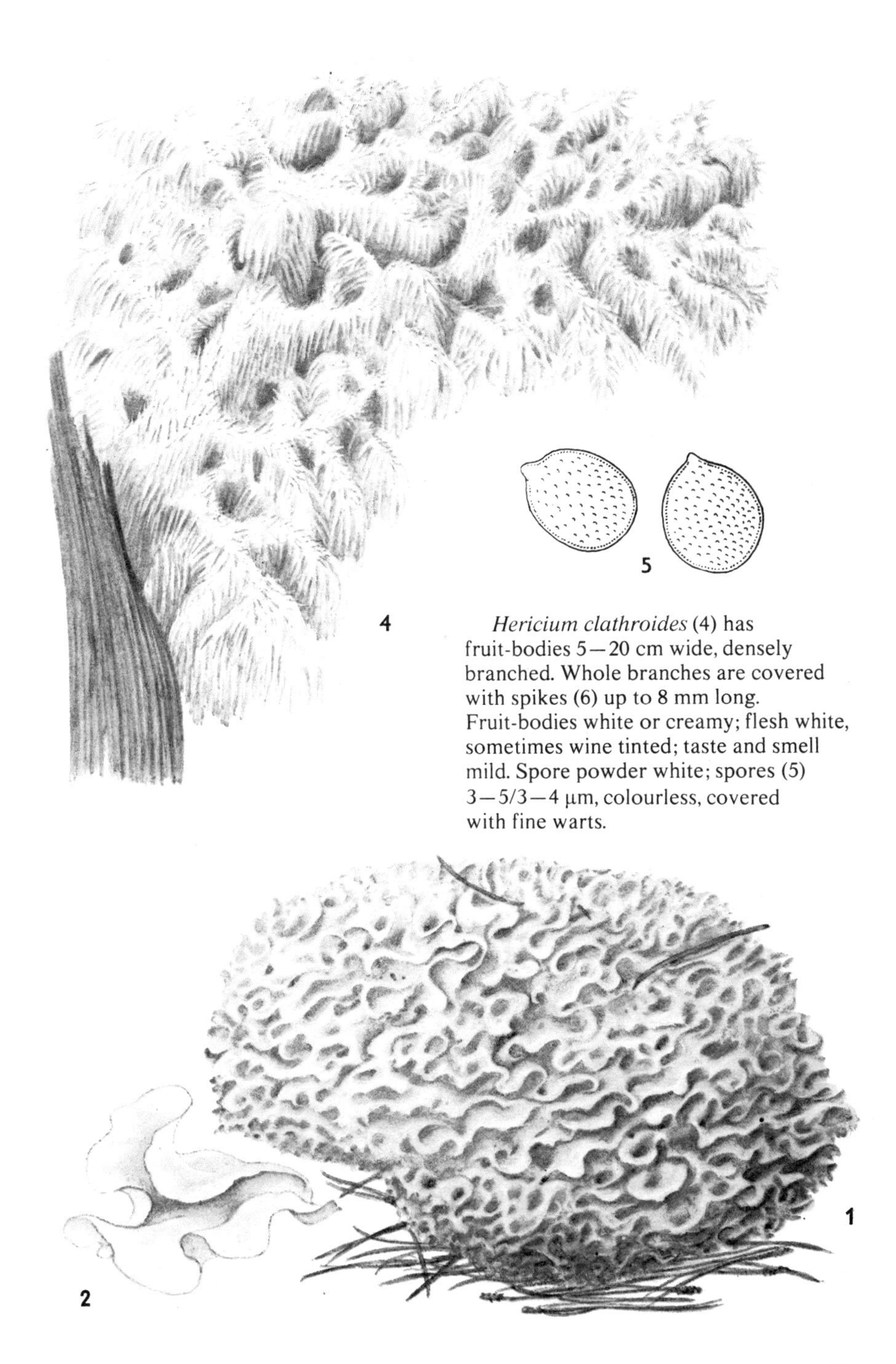


*Hericium clathroides* (4) has fruit-bodies 5—20 cm wide, densely branched. Whole branches are covered with spikes (6) up to 8 mm long. Fruit-bodies white or creamy; flesh white, sometimes wine tinted; taste and smell mild. Spore powder white; spores (5) 3—5/3—4 µm, colourless, covered with fine warts.

## Wood-hedgehog, Hedgehog Mushroom

Tooth Fungi
*Hydnaceae*

*Hydnum repandum* L. ex FR.
Syn.: *Dentinum repandum*

The Tooth Fungi differ from all the other Basidiomycetes by the presence of spines on the lower surface of the cap. The spines have the same function as the gills or tubes; they bear the hymenium on their surface. The best known genera of this family are *Hericium sarcodon* and *Hydnum* with the species *H. repandum,* which grows abundantly throughout the temperate zone of the northern and southern hemispheres. The Wood-hedgehog thrives in summer and autumn in deciduous, coniferous and mixed forests from lowlands to highlands. It lives in mycorrhizal association with beeches and spruces and prefers clay and lime soils. Its fruit-bodies grow individually rarely; most often they form clusters, strips or circles. They are edible but because they are tough and rather sour, they are best suited for use in mixtures with other mushrooms. The colour and shape of the Wood-hedgehog vary a great deal and individual mushrooms may differ very much from the illustrated specimen; the most striking form has pure white fruit-bodies. Beige fruit-bodies resemble a terrestrial polypore, *Albatrellus confluens,* which has small tubes on the lower surface of the cap.

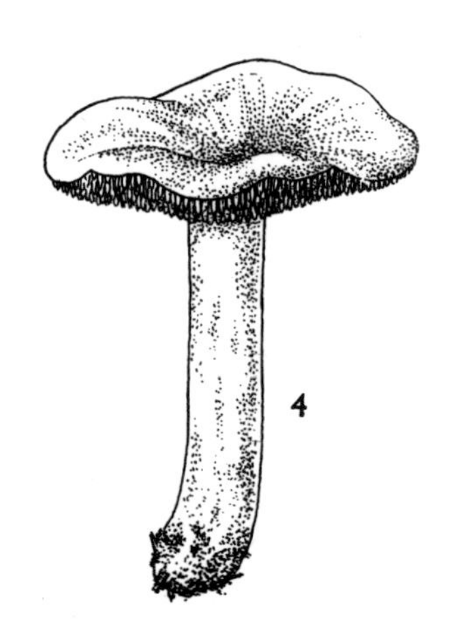


*Hydnum repandum* has cap 6—15 cm wide, contorted, firm but fragile, dry, downy but later becoming bare, whitish to beige, yellowish. Spines 2—6 cm long, whitish or yellowish, decurrent to the stipe. Stipe 2—7/1—3 cm, usually eccentric, white or yellowish, downy, later bald. Fragile flesh white or yellowish (1); taste slightly acid, aroma pleasant. Spore powder whitish; 6.5—9/5.5—7 μm, colourless, containing small grains (2) or a large droplet (3). The Wood-hedgehog can be confused with the closely related, edible *H. rufescens* (4), which has smaller fruit-bodies; its cap is 7 cm in diameter at maximum, stipe is central, spines are not decurrent and its overall colour is darker. It grows in the same habitats as the Wood-hedgehog.

1
2
3

*Albatrellus confluens*
(ALB. et SCHW. ex FR.) KOTL. et POUZ.
Syn.: *Polyporus confluens*

Polypores
*Polyporaceae*

The Aphylophorales form an important group of fungi which aid the decomposition of wood in forests. They either grow as wood saprophytes on dead trunks, stumps or branches or as parasites on living trees. The genus *Albatrellus* is an exception, as all its 19 species found in the northern hemisphere are terrestrial mushrooms, living as saprophytes or forming a mycorrhizal association with trees. During their growth *Albatrellus* fruit-bodies, unlike the majority of Polypores, are not capable of absorbing small stalks and branches which are in their way. Therefore they grow in the same way as Boleti or Gill Fungi, simply pushing obstacles aside.

The commonest of six European species are *Albatrellus confluens* and *A. ovinus*. They grow in coniferous forests, mostly in acid soils of submountainous regions. The fruit bodies of *A. ovinus* develop separately and exclusively under spruce trees, while the fruit-bodies of *A. confluens* form clusters under various coniferous species. The most favourable conditions for the formation of fruit-bodies are in summer and autumn. The closely related *A. subrubescens* can be found under pines and firs, but is distinguished by its bitterish taste. Deciduous forests, especially beech growths, host *A. cristatus*, which has a brown-green or olive cap. All the above-mentioned *Albatrellus* species are edible, but not easily digested due to their toughness.

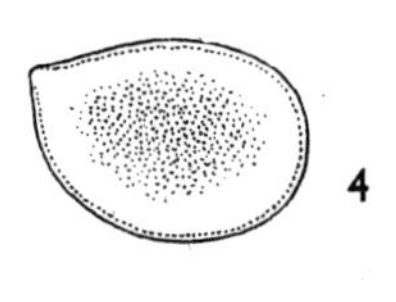


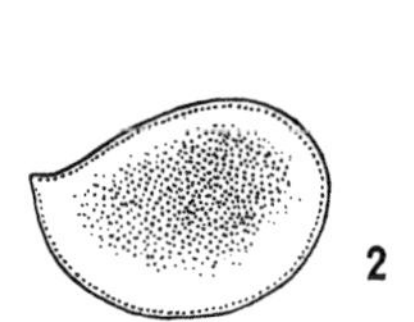


*Albatrellus confluens* (1) has cap 3—8 cm wide, ochre or beige, with minute white tubes. Stipe is 2—8/1—4 cm, white. Flesh is white, but a drop of ferrous sulphate colours it brick red. Taste is bitterish, smell indistinct. Colourless spores (2) are 4—5/3—3.5 μm.

*Albatrellus ovinus* (3) has cap 3—15 cm wide, whitish or yellowish, often cracked. The minute tubes are white or yellowish, becoming yellow when bruised. Stipe is 2—8/1—4 cm, white. Flesh is white turning yellow; a drop of ferrous sulphate colours it grey. Taste pleasant, smell indistinct. Spores (4) are 3.5—4.4/3—4 μm, and are colourless.

1

3

## Birch Bracket

Polypores

*Piptoporus betulinus* (BULL. ex FR.) KARST.

*Polyporaceae*

Fruit-bodies of this mushroom appear on dead birch trunks and branches and infect living birches, usually where the trunks have been damaged, or the branches broken off, or in cracks caused by frost. The Birch Bracket is an annual, its fruit-bodies die off the same year, or last until the following spring. The mycelium, which penetrates the wood, bears new fruit-bodies. The infection usually spreads from the top of the trunk to its base or from the ends of the branches towards the trunk. The mycelium causes red-brown rot; the rotten wood can be easily rubbed into a fine red-brown powder with the fingers. This is used in Switzerland for polishing watch parts. The Birch Bracket is widely distributed in birch habitats through the temperate zone of the northern hemisphere, particularly in Germany, Russia and North America. A related, but much rarer species is *P. pseudobetulinus* with thinner and paler fruit-bodies and with a sharp edge of the cap which is never involuted. It grows on the trunks of aspen trees. The similar *Buglossoporus quercinus,* which attacks living oaks, also has fruit-bodies with a sharp edge wich does not turn under; it has large pores, 1.5—3 pores to 1 mm.

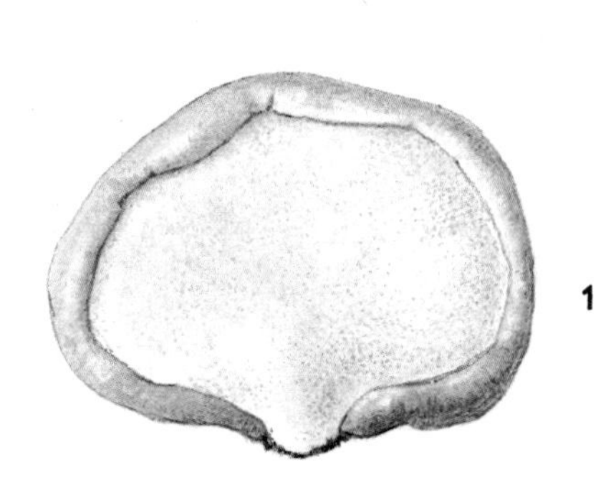


*Piptoporus betulinus* has hoof-shaped cap 5—20 cm wide, 2—7 cm high, with thin and smooth, paper-like, grey-brown or brown cuticle which cracks later. The involuted edge (1) forms a rise along the entire perimeter of the fruit-body; tubes arranged in one layer, thin and whitish. Very small circular pores (3—6 pores to 1 mm) whitish, later with ochre tint. Flesh (2) white, spongy and soft, later tough and cork-like; taste and smell sour. Spores (3) colourless, 4.5—6/1.3—2 μm. Fruit-bodies consist of two types of hyphae, thin-walled with clamps and thick-walled or completely filled hyphae.

3

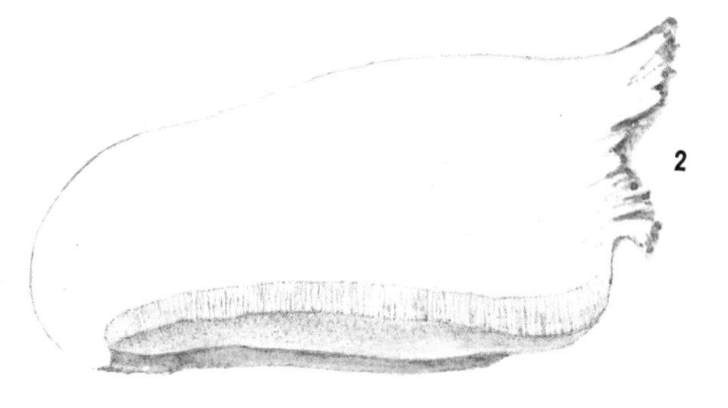
2

## The Chanterelle

*Cantharellaceae*

*Cantharellus cibarius* FR.

The Chanterelle, well-known in Europe, North and South America, Africa, China and Japan, grows in all types of forest in summer and in autumn, forming clusters, strips and circles, always containing a large number of fruit-bodies. It forms mycorrhiza with various trees, including pine, spruce, oak and elm. The raw fruit-bodies have a pungent taste which disappears with cooking. Their toughness makes them somewhat indigestible to sensitive stomachs. The fruit-bodies resist attack by insects and therefore last long in the wild. Chanterelles are very variable in shape and colour. The variety *albus* has pure white fruit-bodies; var. *albipes* has a white stipe only; var. *amethysteus* has a cap covered with red-purple scales. The underside is covered with gill-like folds.

False Chanterelle, *Hygrophoropsis aurantiaca* (fam. Paxillaceae) has well-developed gills, which often branch like a fork. Its growing season coincides with the Chanterelle. It is edible, but not very good.

*C. tubaeformis* grows in autumn most often in coniferous forests together with the related *C. lutescens*, which has a yellow-brown cap with orange-tinted folds, and *C. infundibuliformis*, which has a dark brown cap and a yellow stipe. All species are edible.

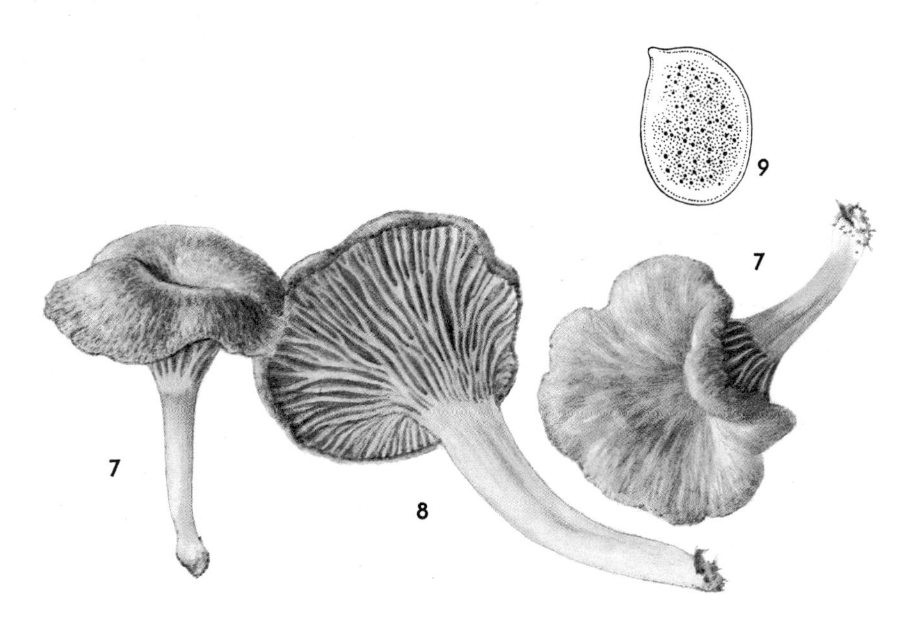

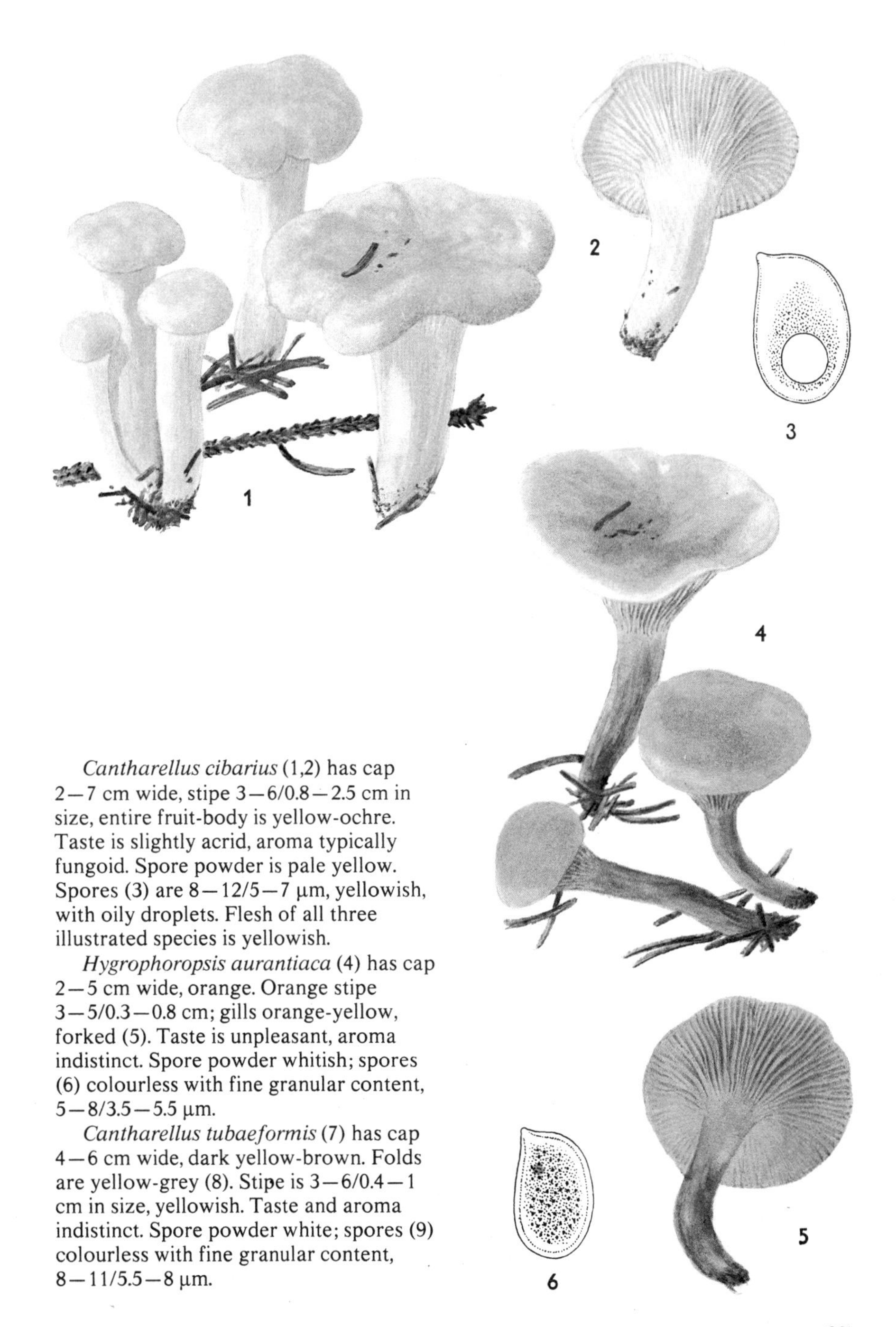


*Cantharellus cibarius* (1,2) has cap 2—7 cm wide, stipe 3—6/0.8—2.5 cm in size, entire fruit-body is yellow-ochre. Taste is slightly acrid, aroma typically fungoid. Spore powder is pale yellow. Spores (3) are 8—12/5—7 µm, yellowish, with oily droplets. Flesh of all three illustrated species is yellowish.

*Hygrophoropsis aurantiaca* (4) has cap 2—5 cm wide, orange. Orange stipe 3—5/0.3—0.8 cm; gills orange-yellow, forked (5). Taste is unpleasant, aroma indistinct. Spore powder whitish; spores (6) colourless with fine granular content, 5—8/3.5—5.5 µm.

*Cantharellus tubaeformis* (7) has cap 4—6 cm wide, dark yellow-brown. Folds are yellow-grey (8). Stipe is 3—6/0.4—1 cm in size, yellowish. Taste and aroma indistinct. Spore powder white; spores (9) colourless with fine granular content, 8—11/5.5—8 µm.

## Horn of Plenty, Trumpet of the Dead *Cantharellaceae*

*Craterellus cornucopioides* (L. ex FR.) PERS.

The characteristic features of the unsightly but interesting Horn of Plenty are its black colouring, funnel-like shape and clustery growth. Clusters of its fruit-bodies in various stages of development can be found in autumn under the fallen leaves of oaks and beeches; they range from small regular funnels to large, hollow fruit-bodies with contorted, folded and torn margins. Despite their uninviting appearance they have a good and slightly spicy flavour. The Horn of Plenty is widespread in the temperate zone of the northern hemisphere and probably also in tropical regions. It forms mycorrhiza in deciduous trees, particularly oak and beech. The species can be confused with *Cantharellus cinereus,* which grows in sparse clusters in deciduous forests. Its characteristic features are its fruity aroma, grey-black colouring and particularly well-developed veins and folds on the lower surface of the cap. Another related species, *Pseudocraterellus sinuosus,* has grey and grey-brown fruit-bodies; the undulate lower surface of the cap is light grey with a yellow-ochre tinge. It is a relatively rare inhabitant of deciduous forests, where it grows in groups in acid soil. Both species appear in summer and autumn and are edible.

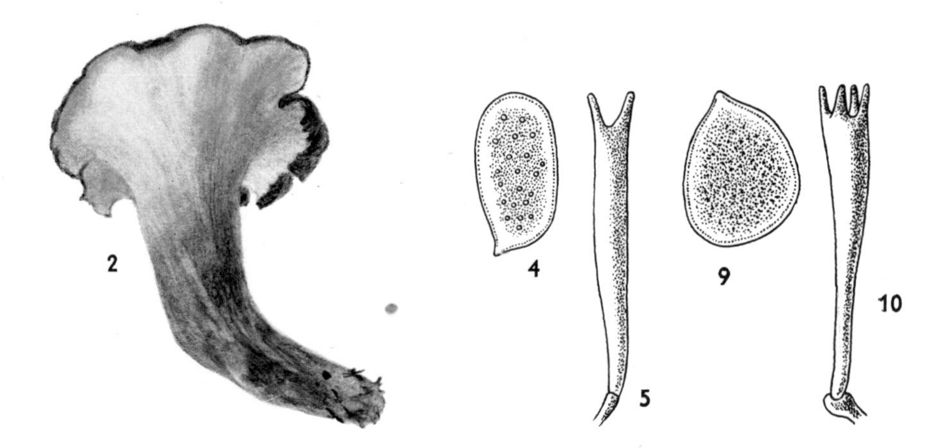

*Craterellus cornucopioides* (1) has cap 3—8 cm wide, brown-black or black on the surface. Stipe 5—12/1—2 cm. The smooth lower surface of the cap and the stipe are pruinose (2). Flesh grey-black or black (3); taste and aroma are pleasant. Spore powder white; spores (4) colourless, 10—17/6—10 µm. Basidia often have two stalks (5).

*Cantharellus cinereus* (6) has cap 2—6 cm wide. Stipe is 3—8/0.5—1.5 cm in size. Cap and stipe are grey-black, folds are ash grey with whitish bloom (7). Flesh is grey-brown, later turning black (8); taste nondescript, aroma strong and fruity. Spore powder white; spores (9) colourless, 7—14/5—8,5 µm. Basidia have four stalks (10).

## Oyster Fungus

*Pleurotaceae*

*Pleurotus ostreatus* (JACQ. ex FR.) KUMM.

The Oyster Fungus is a true winter mushroom. Its fruit-bodies appear soon after the first frost and then through winter until spring. It clusters in overlapping tiers on deciduous trees, including beeches, willows, poplars and birches, in gardens, parks and forests in all temperate regions. The Oyster Fungus is very popular for its excellent flavour. It can be prepared and cooked in any way; it is never attacked by parasites and grows when there are no other wild mushrooms available. The Oyster Fungus is successfully cultivated both in its natural habitat, where dead stumps, trunks and logs are injected with the mycelium, and in nurseries on cut straw or crushed corn husks. The Oyster Fungus can be confused with several closely-related species, particularly *P. columbinus,* which forms its fruit-bodies in autumn till the first frost and which has a blue-grey to purple-brown cap and gills reduced to veins which descend down the stipe. It grows on coniferous and rarely also on deciduous trees.

*P. pulmonarius* has characteristic fan-shaped fruit-bodies; its gills and the margin of the cap turn yellow. It can be found on deciduous trees from spring to autumn. Edible.

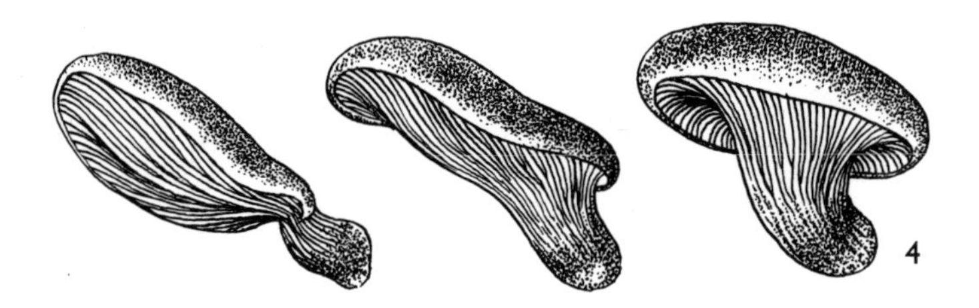


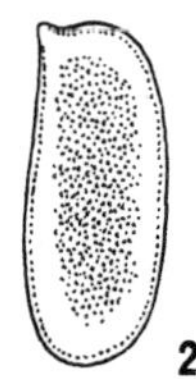


*Pleurotus ostreatus* (1) has cap 5—20 cm wide, shell-shaped, dark brown, black-grey, blue-purple; decurrent gills white or brownish; stipe 2—4/1—2 cm, whitish, attached to the cap in different ways. Flesh is white, taste and aroma pleasantly fungoid. Spore powder white to brownish; spores (2) colourless, 7.5—12/3—4 μm.

*Pleurotus pulmonarius* (3) has cap 5—25 cm wide and stipe 2—3/0.6—1.5 cm in size, fan-shaped cap and stipe are white or pale hazel brown. Gills are whitish or creamy ochre. Flesh is white. Entire fruit-body turns yellow, especially when old. Taste is inconspicuous, smells faintly of urine. Spore powder white or creamy; spores (2) colourless, 7.5—11/3.2—4 µm.

# *Hygrophorus pustulatus* (PERS. ex FR.) FR. *Hygrophoraceae*
Syn.: *Limacium pustulatum*

*Hygrophorus* fruit-bodies do not reach large sizes, but are attractively coloured in bright red, green and orange. About 40 species grow in Europe, where they form mycorrhizal associations predominantly with coniferous, but also with deciduous trees. Many species can also be found outside forests, in meadows or pasturelands from spring until autumn. Most of them have slimy caps and sometimes also stipes. During damp or sunny weather the caps are vitreous and luminous. Though the following species are edible, their poor, watery flavour makes them best for mixing with other, tastier species.

*Hygrophorus pustulatus* is an autumn species, which grows until the first frost. It is particularly abundant in mountainous spruce forests, but it can be also found in deciduous and mixed woods. Its characteristic features include a slimy grey-brown cap and a grey-black granular stipe.

Larch Hygrophorus, *Hygrophorus lucorum,* is dependent for its existence on the larch, which it follows high up into the mountains. Its fruit-bodies grow gregariously in clusters, strips and circles in autumn only, until the beginning of winter. The conspicuous *H. speciosus* grows also in mycorrhiza with larches in mountainous areas. It has a chrome yellow cap with an orange umbo and white striate, yellow stipe.

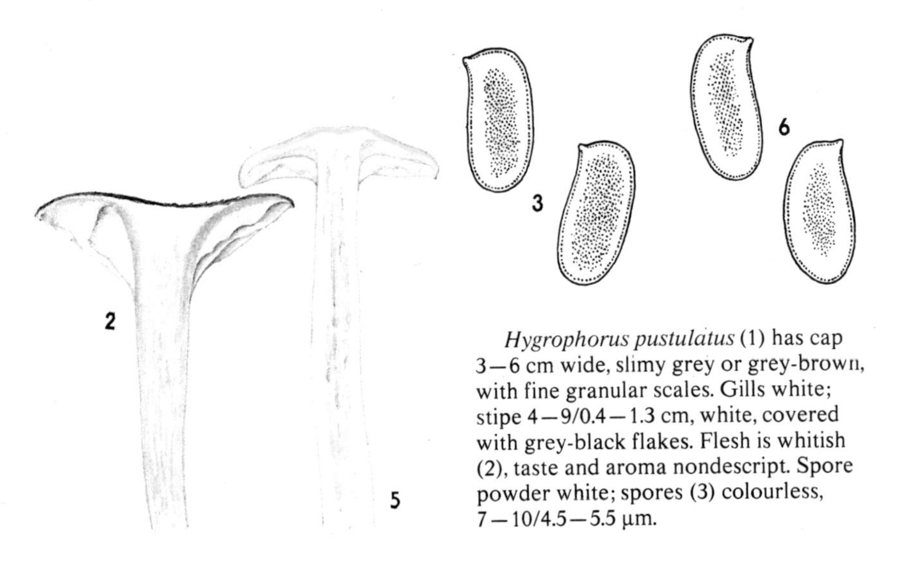


*Hygrophorus pustulatus* (1) has cap 3—6 cm wide, slimy grey or grey-brown, with fine granular scales. Gills white; stipe 4—9/0.4—1.3 cm, white, covered with grey-black flakes. Flesh is whitish (2), taste and aroma nondescript. Spore powder white; spores (3) colourless, 7—10/4.5—5.5 µm.

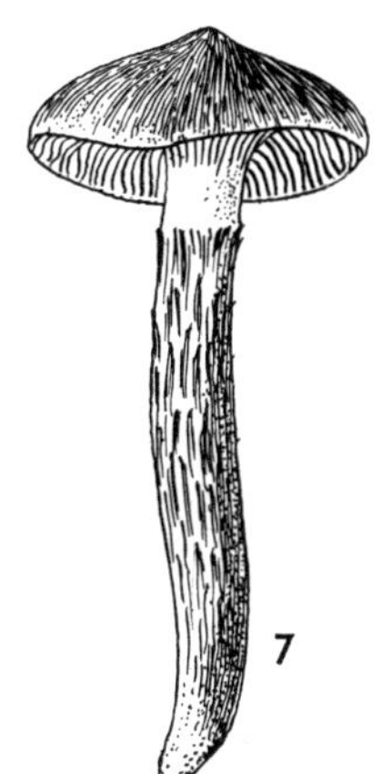


*Hygrophorus lucorum* (4) has cap 2—6 cm wide, smooth, slimy, lemon yellow, darker in the centre, gills whitish, sometimes pale yellow; stipe 4—9/0.4—1.3 cm in size, slimy, with flakes, whitish or yellowish. Flesh whitish (5), taste and aroma nondescript. Spore powder white; spores (6) colourless, 7—10/4—6 μm.

*Hygrophorus olivaceoalbus* can be found in similar habitats as *Hygrophorus pustulatus.* It differs by its olive brown transversely striate stipe (7).

## Spring Hygrophorus

*Hygrophoraceae*

*Hygrophorus marzuolus* (FR.) BRES.

*Hygrophorus marzuolus* is irregularly distributed in central and southern Europe, North America and northern Africa. The fruit-bodies occur in spring from March until June, and in southern Europe from January. *H. marzuolus* is characteristic of natural growths of mountainous coniferous forests. Sometimes it even reaches the lowlands, where it inhabits lime-rich soils. It lives in mycorrhiza with spruces, pines and firs and in Morocco it grows under cedar trees almost 2,000 m above sea level. The fruit-bodies are usually concealed by fallen needles for a long time, or even under the melting snow. As soon as one fruit-body is found, many others can be discovered under the piles of needles in its neighbourhood.

*Hygrophorus marzuolus* closely resembles *H. camarophyllus*, which is sometimes considered to be its autumnal form. While *H. marzuolus* has a contorted, almost white or steel grey cap with a greyish crooked stipe, *H. camarophyllus* has a brown cap and stipe and the cap has radiating fibrils. It grows under pines and spruces in acid soil. In Europe it can be found only in autumn, but in North America it appears at the beginning of spring.

*Hygrophorus russula* is also quite rare. Its fruit-bodies appear in autumn, and sometimes as late as December, particularly in southern Europe. For its growth it needs a limey soil and warm deciduous forests. It forms mycorrhizal associations with oaks. All these above-mentioned species are edible and have a good flavour.

*Hygrophorus marzuolus* (1) has cap 5—10 cm wide, dry, wavy, pruinose, whitish greyish or dark steel grey. Gills shortly decurrent down the stipe, whitish, ash-grey or grey when old. Stipe 5—8/0.6—2.5 cm, rather crooked, with silky fibrils; whitish, with grey tinge when old. Flesh white, slightly silvery when old. Spore powder white; spores (2) colourless, 7—9/4—5 µm.

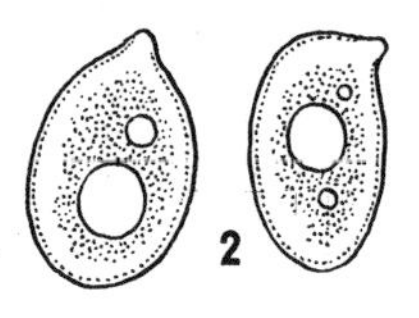

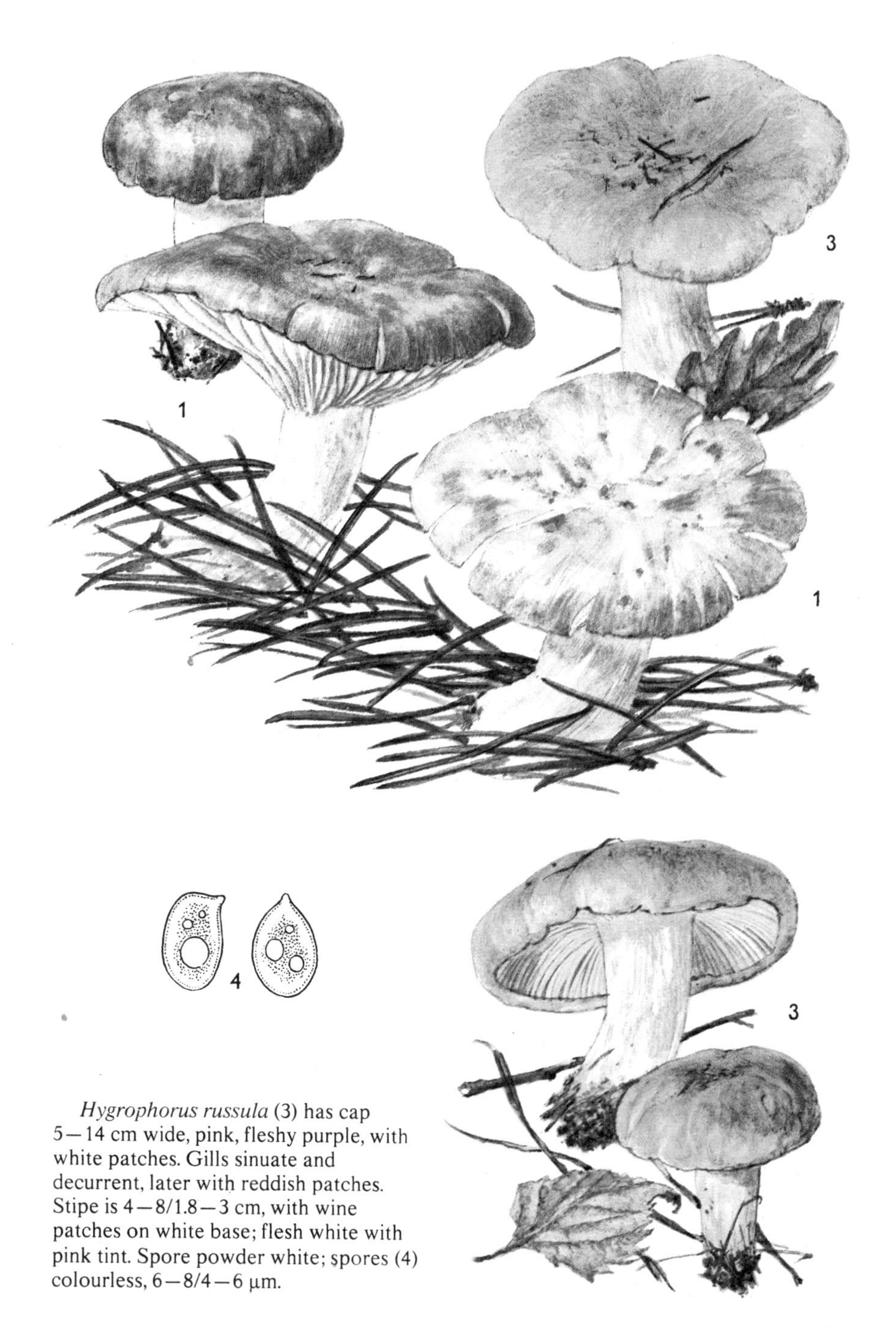


*Hygrophorus russula* (3) has cap 5 – 14 cm wide, pink, fleshy purple, with white patches. Gills sinuate and decurrent, later with reddish patches. Stipe is 4 – 8/1.8 – 3 cm, with wine patches on white base; flesh white with pink tint. Spore powder white; spores (4) colourless, 6 – 8/4 – 6 µm.

## St George's Mushroom

*Tricholomataceae*

*Calocybe gambosa* (FR.) SING.
Syn.: *Tricholoma gambosum*

The genus *Calocybe* has been separated from the genus *Tricholoma* because of the differences in the anatomy of their fruit-bodies. The genus includes about 20 species, of which the St George's Mushroom, a typical spring mushroom which flourishes outside forests, is the most common and best-known. It can be found in clusters, strips and circles in the grass of orchards, on hillsides, in parks, gardens and along the edges of forests from lowlands to highlands. It probably lives in mycorrhizal association with shrubs and trees belonging to the order Rosales, such as roses, blackthorns, plums and hawthorns. In northern Africa it grows under oak shrubbery. Typical colouring of fruit-bodies is white; their taste and aroma are strongly floury.

Anatomically and morphologically similar fruit-bodies, straw yellow or yellow-ochre in colour, can be found in summer and autumn and sometimes also in spring. Fruit-bodies of this colouring often inhabit deciduous forests and are either considered to be a variety of St George's Mushroom *(C. gambosa* var. *flavida)* or a separate species, *C. georgii.* Both species have a pleasant taste and aroma. Careless collectors can confuse them with several similar species, of which *Inocybe patouillardii* and *Entoloma sinuatum* are strongly poisonous.

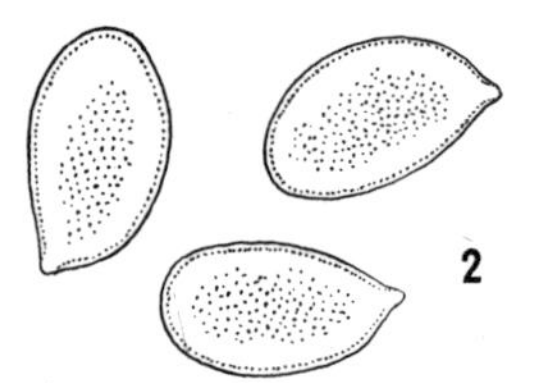


*Calocybe gambosa* (1) has cap 2.5—8 cm wide and stipe 5—8/1—2 cm in size; cap, stipe and gills are whitish, creamy and smooth. Flesh is white, taste and aroma floury. Spore powder white; spores (2) colourless, 4—6/2—3.5 μm.

*Calocybe georgii* (3) has cap, stipe and gills straw yellow or yellow-ochre, other features are the same as in *C. gambosa.*

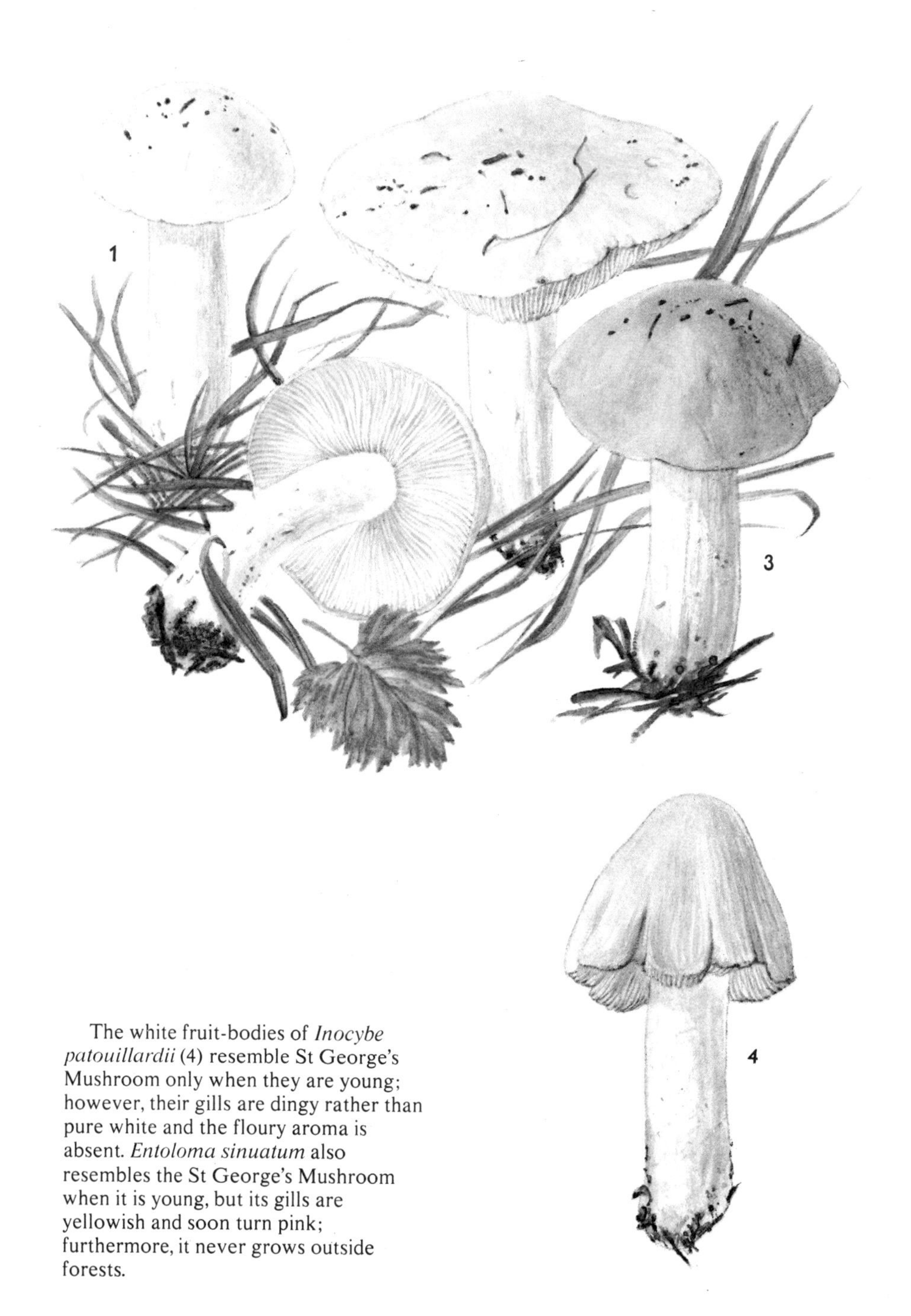


The white fruit-bodies of *Inocybe patouillardii* (4) resemble St George's Mushroom only when they are young; however, their gills are dingy rather than pure white and the floury aroma is absent. *Entoloma sinuatum* also resembles the St George's Mushroom when it is young, but its gills are yellowish and soon turn pink; furthermore, it never grows outside forests.

## **Deceiver, Common Laccaria** *Tricholomataceae*

*Laccaria laccata* (SCOP. ex FR.) BK. et BR.

Members of the genus *Laccaria* are all coloured purple, pink or orange. The gills are also of the same colour, but the spore powder is white, sometimes with a lilac tint. The spores are colourless and their surface is spiny (with the exception of *L. maritima).* These mushrooms are saprophytes and occasionally live in mycorrhizal association with deciduous trees. Their colouring and size are very variable: in damp weather the colour is deep and sharp, while in dry conditions the caps and the stipe are almost white. Some *Laccaria* species are narrowly linked to a certain habitat. *L. proxima* grows predominantly in moss on peat-bogs; *L. maritima* in sand of seaside dunes; *L. montana* is an alpine and subalpine species; and *L. tortilis* can be found on bare soil in thickets along tracks.

The three illustrated species are the most abundant representatives of their genus and can be found in the entire temperate zone of the northern hemisphere. They are all edible. *L. proxima* is closely related to *L. laccata,* but it differs by its squamous cap, long scabrously fibrilous stipe and by its habitat. *L. laccata* grows profusely in summer and autumn in all types of forests, reaching high into the mountains to 3,000 m above sea level. Purple Laccaria, *L. amethystina,* differs from the previous species by its beautiful purple fruit-bodies. It forms clusters in coniferous and deciduous forests from June to November.

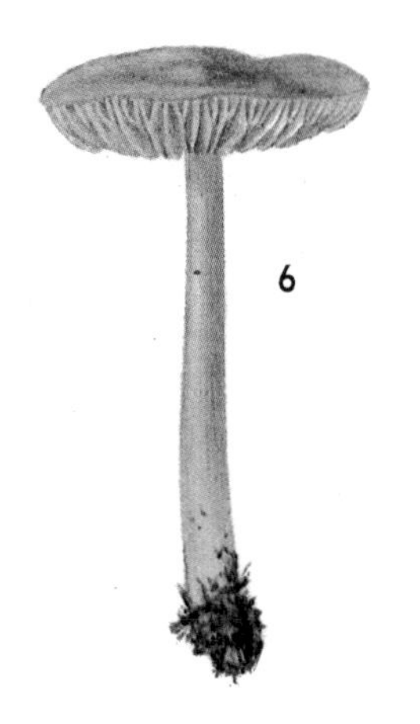


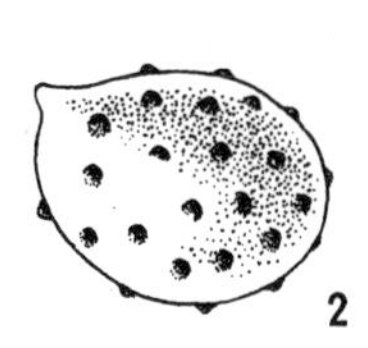


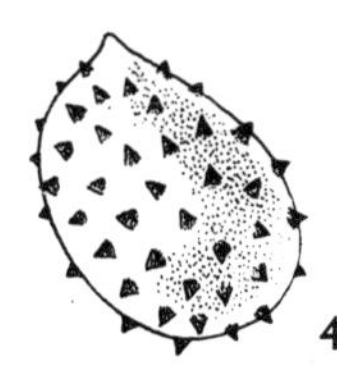


*Laccaria laccata* (1) has cap 1—5 cm wide, stipe is 3—8/0.2—0.6 cm in size; cap and stipe are pink, pink-brown or ochre, fibrilous; gills and flesh are also pink. Spores (2) 7—9/6—7.5 µm.

*Laccaria proxima* (3) has cap 1–7 cm wide, rusty, hygrophane, squamous; gills are pink. Stipe 3–12/0.3–1 cm, orange-brown, with rough brownish fibrils. Spores (4) are 7.5–10.5/6.7–7.6 µm.

*Laccaria amethystina* (5) has cap 2–6 cm wide, stipe 2.5–10/0.5–0.8 cm; entire fruit-body is purple or purple-blue, becoming paler in drought. Spores are 8–9.5 µm long. *L. amethystina* resembles *Mycena pura* (6). Taste and aroma of all three species are nondescript.

## *Lepista inversa* (SCOP. ex FR.) PAT. — *Tricholomataceae*

Syn.: *Clitocybe inversa*

*Clitocybe* is among the richest European genera, containing 100 species. They are very difficult to classify because their colouring and microscopic features are very similar. Some species are edible but their quality is inferior; on others little further information is available; still other species are strongly poisonous. These are usually white, such as *Clitocybe cerussata* or *C. dealbata.* The majority of species are widespread throughout the temperate zone of the northern hemisphere. *Clitocybe* species are typical humus saprophytes; their wooly mycelium thoroughly penetrates the top layers of the soil, which it decomposes with its enzymes. Through this decomposition these mushrooms acquire the carbonaceous substances necessary for their nourishment and growth.

*Lepista (Clitocybe) inversa* can be found in autumn in spruce thickets where it forms clusters, strips and circles, containing tens or hundreds of fruit-bodies. It is characterized by a red-yellow cap and stipe and yellow-rusty gills. The similar *Lepista gilva* is ochre with yellowish gills. In autumn it inhabits coniferous and also deciduous forests. Both species, once considered to be edible, contain poisonous substances similar to muscarine. *Clitocybe vibecina,* common autumnal mushroom in coniferous forests, particularly of cultivated spruce, is entirely grey; it has an unpleasant musty smell and is inedible.

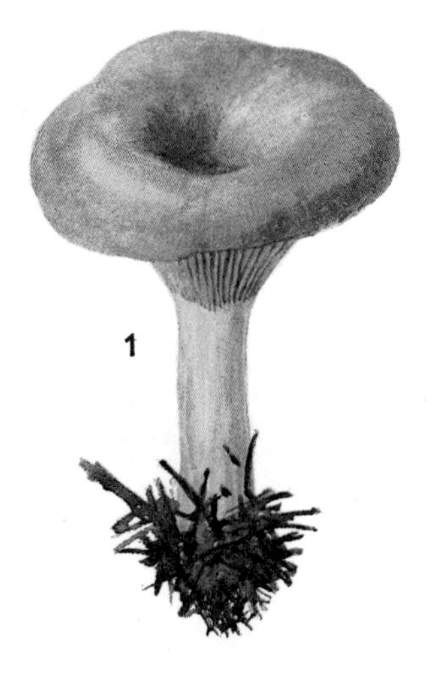


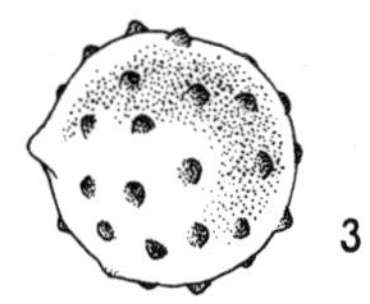


*Lepista inversa* (1) has cap 4—8 cm wide, hygrophane, red-yellow, rusty or with rusty patches. Gills are pale yellowish to rusty; stipe is 3—5/0.4—0.6 cm, same colour as cap. Flesh is ochre (2). Spores (3) are colourless, 3—4 μm in diameter.

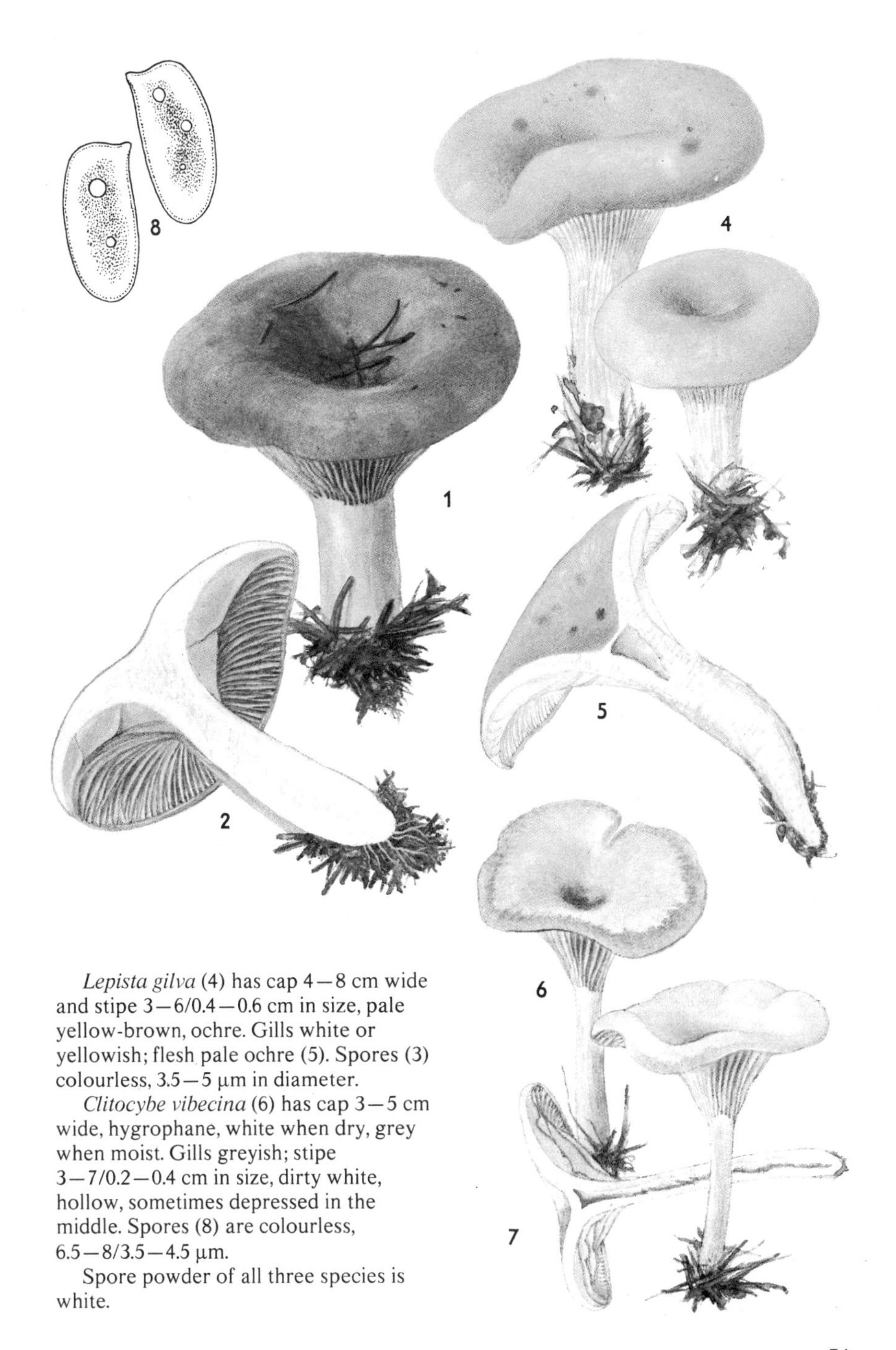


*Lepista gilva* (4) has cap 4—8 cm wide and stipe 3—6/0.4—0.6 cm in size, pale yellow-brown, ochre. Gills white or yellowish; flesh pale ochre (5). Spores (3) colourless, 3.5—5 μm in diameter.

*Clitocybe vibecina* (6) has cap 3—5 cm wide, hygrophane, white when dry, grey when moist. Gills greyish; stipe 3—7/0.2—0.4 cm in size, dirty white, hollow, sometimes depressed in the middle. Spores (8) are colourless, 6.5—8/3.5—4.5 μm.

Spore powder of all three species is white.

# Clouded Agaric

*Tricholomataceae*

*Lepista nebularis* (FR.) HARM.
Syn.: *Clitocybe nebularis*

*Lepista nebularis* belongs to the largest species of this genus. Its cap reaches 20 cm in diameter and the stipe can be 15 cm long. It occurs in large numbers late in autumn in all types of lowland and highland forest; its distribution is worldwide. Its characteristic features include fleshy, ash-grey, aromatic fruit-bodies and a slightly variable shape. It used to be collected in large quantities for eating, despite its strong aroma, but recently opinions of its edibility have changed, since it has caused several cases of poisoning. It is definitely poisonous when raw and when cooked can cause digestive disorders. Its poisoning is characterized by sweating and by discomfort of the respirating and digestive tracts. Fruit-bodies of *Lepista nebularis* also contain heavy metals, namely mercury and cadmium, because this mushroom belongs to the humus saprophytes which absorb these elements from the soil and accumulate them in their fruit-bodies.

Related *Clitocybe clavipes* is smaller and differs by its swollen, club-shaped grey stipe. It grows scattered in coniferous forests from August until October. It is edible, but not very pleasant in taste. Swollen stipes are usually soft and soaked with water when it rains.

*Lepista nebularis* (1,2) has cap usually 7—15 cm wide, ash grey, grey-brown, pruinose. Gills shortly decurrent, whitish or yellowish; stipe usually 6—10/1.5—3 cm, whitish or greyish. Flesh white and aromatic; taste slighty acid. Spore powder is creamy; spores (3) colourless, 6—8/3—4 μm.

*Lepista nebularis* can be confused with the strongly poisonous *Entoloma sinuatum,* particularly with its young fruit-bodies (4). However the cap of *E. sinuatum* lacks grey colouring; it is whitish, yellowish or ochre and its fruit-bodies have no aroma. Gills of *E. sinuatum* become pink in adult fruit-bodies, while in *L. nebularis* they are constant whitish or yellowish.

*Clitocybe clavipes* (5,6) has cap 4—6 cm wide, brown to grey-brown. Gills whitish or yellowish; stipe spindle-shaped, brown-grey, consists of long fibres. Flesh is white; taste and aroma slightly reminiscent of bitter almonds. Spore powder white; spores colourless, 5—7/3—4 μm. Gills of *Lepista* species are decurrent (7).

## Wood Blewit, Naked Mushroom

*Tricholomataceae*

*Lepista nuda* (BULL. ex FR.) CKE.
Syn.: *Tricholoma nudum*

The genus *Lepista* includes only a few species, all resembling mushrooms of the genera *Tricholoma* and *Clitocybe.* They differ from them mostly by certain microscopic features. They are humus saprophytes widespread in Europe, north Africa, Asia and North America. *Lepista nuda* and *L. personata* are large and fleshy autumnal species, but due to their purple colouring they are not very popular with mushroom pickers. If well cooked, however, they are edible and have a delicious flavour.

*Lepista nuda* grows gregariously in the humus of deciduous and occasionally coniferous forests from September to December, if frosts are not too severe. Individual fruit-bodies appear even in spring. Experiments with artificial cultivation have been successful, since the fruit-bodies develop on natural substrates consisting of leaves and needles. In future this mushroom could be cultivated in bulk as with Field Mushrooms and Oyster Fungi.

*Lepista personata* is a robust mushroom in which the cap and stipe are differently coloured. This compact mushroom is usually found in meadows, pastures, gardens or at the edges of forests; sometimes it also grows in deciduous woods. *L. sordida* has half-sized grey-purple or brown-purple fruit-bodies and grows on compost heaps, in fields and pastures. It is also edible.

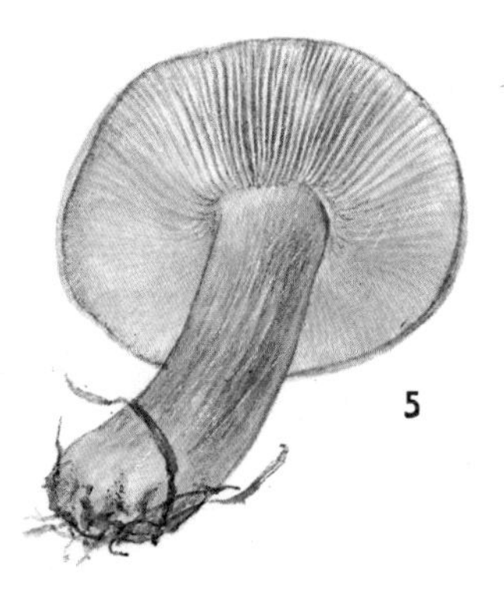


*Lepista nuda* (1) has cap 6—15 cm wide, purple, brown-purple, brownish in the centre. Gills are purple; stipe 5—18/1—2.5 cm, purple and filamentous. Flesh purple or greyish (2); taste and aroma are distinct, as if perfumed or resembling radish.

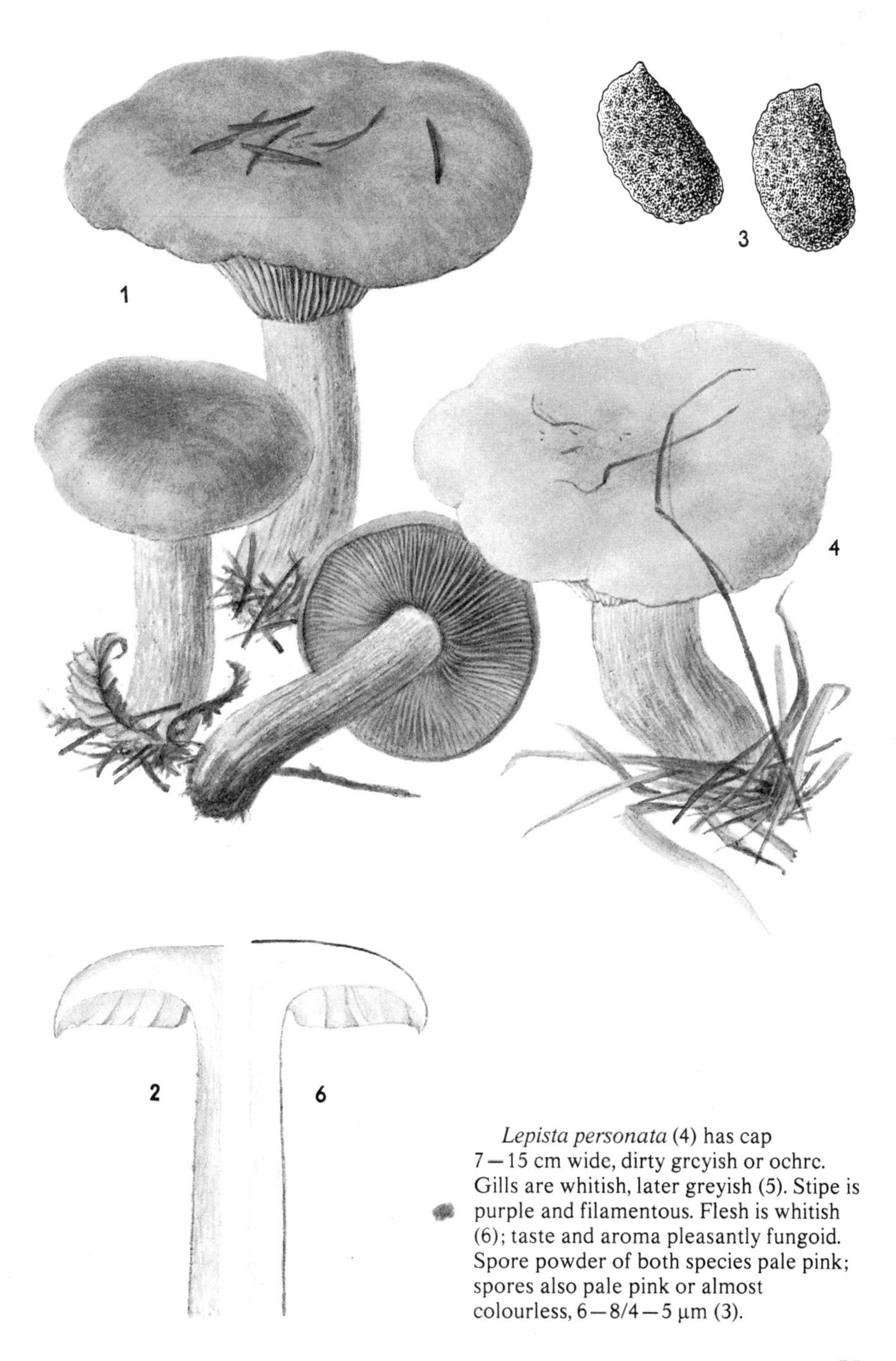


*Lepista personata* (4) has cap 7–15 cm wide, dirty greyish or ochre. Gills are whitish, later greyish (5). Stipe is purple and filamentous. Flesh is whitish (6); taste and aroma pleasantly fungoid. Spore powder of both species pale pink; spores also pale pink or almost colourless, 6–8/4–5 μm (3).

## **Honey Fungus Boot-lace Fungus** *Tricholomataceae*

*Armillariella polymyces* (PERS. ex S. F. GRAY) SING. et CLC.
Syn.: *Armillaria mellea* s. l.

The Honey Fungus is distributed almost all over the world. It grows in autumn in clusters on stumps, trunks and roots and also on living trees. In forests it has a great share in the decomposition of stumps and trunks. It is mainly a wood saprophyte which attacks parasitically only weak or old trees. It often attacks spruces, usually in areas which are not their original habitat, and in the course of ten years can damage entire forests. The infection spreads in two ways: by spores, which can infect only dead wood; and by the mycelium, which penetrates even living roots. The mycelium penetrates the soil or passes under the bark with its strong, hard, net-like branching rhizomorphs. The mycelium itself, which spreads through the wood, is pure white and fine. Rotten wood penetrated with mycelium is luminous in the dark; this phenomenon is connected with enzymatic decomposition of the mycelium (chemoluminiscence).

Until recently the Honey Fungus was considered a single but strongly variable species. It has now been divided into several better defined species. For example the Honey Fungus found in June and July in small clusters with pure white spore powder is *A. praecox; A. tabescens* has the same features as *A. mellea,* except for lacking the ring; it grows most frequently on oak trees. The Honey Fungus is poisonous when raw, but is delicious when thoroughly cooked. The lower section of the stipe is tough, almost woody, and is therefore best avoided.

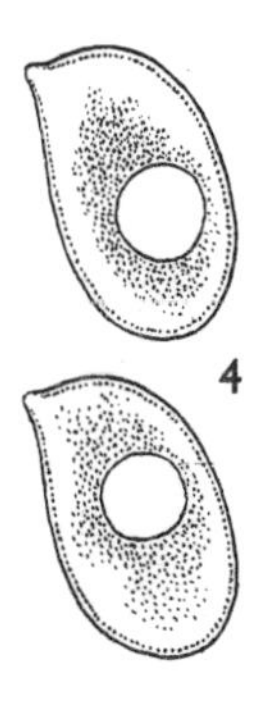


*Armillariella polymyces* (1) has cap 2—10 cm wide, honey-coloured, fleshy or olive-brown, covered with dark brown or red-brown flake-like scales, viscous. Gills whitish, later yellowish or dirty rusty brown, sinuate-decurrent. Stipe 3—15/0.5—2 cm, honey-coloured or fleshy with rusty filaments and whitish ring; flesh is white (2), taste acrid to bitterish, aroma indistinct. Mycelium forms black rhizomorphs (3). Spore powder whitish; spores (4) colourless, 7.5—8.5/5—5.5 µm. Cystidia are in clusters along the edges of gills (5).

1

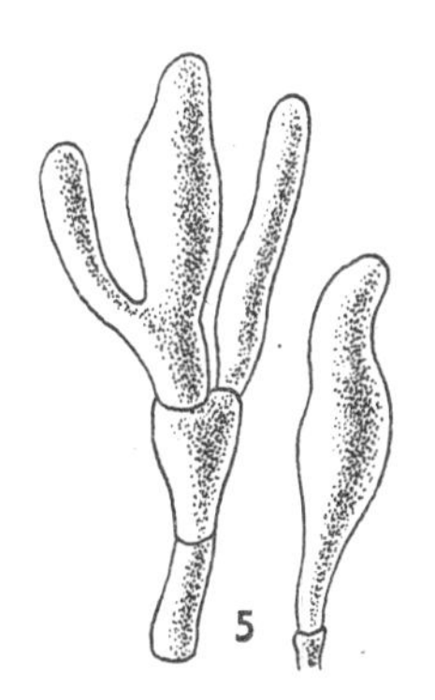
5

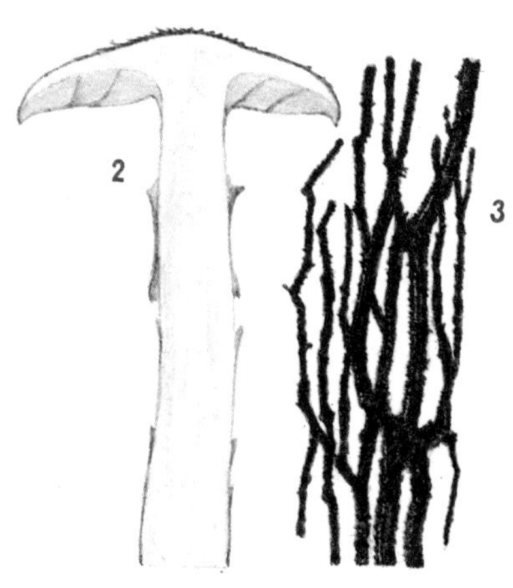
2
3

## **Purple Blewit** *Tricholomataceae*

*Tricholomopsis rutilans* (SCHFF. ex FR.) SING.

The Purple Blewit and *Mycena viscosa* are wood saprophytes. The Purple Blewit is widespread in Europe, North America, Africa and Asia. It grows in summer and autumn in clusters on stumps and dead trunks of coniferous trees, and sometimes out of dead roots close to tree stumps. The Purple Blewit has a variable colouring and shape; its fruit-bodies are sometimes almost yellow, and sometimes purple scales completely cover the original yellow colouring of the cap. The fruit-bodies have a central or eccentric stipe. It is an edible mushroom, but opinions vary about its quality and flavour.

A similar, also edible, species, *Tricholomopsis decora,* which forms clusters on coniferous stumps or dead trunks, has a yellow-orange cap and a stipe with greenish, olive brown scales. It can sometimes be confused with the inedible *Gymnopilus junonius,* which is also yellow or rusty yellow, but has a ring on the stipe. The latter forms sparse clusters on dead and living deciduous trees.

*Mycena viscosa* is a rare species in Europe and other continents. Its viscous, sticky fruit-bodies grow in autumn on dead trunks of coniferous trees or on their stumps. *M. epipterygia,* also viscous, is very abundant. It differs from the previous species by its whitish gills and nondescript taste, and by its habitat in moss of damp coniferous and deciduous forests. Both species are edible.

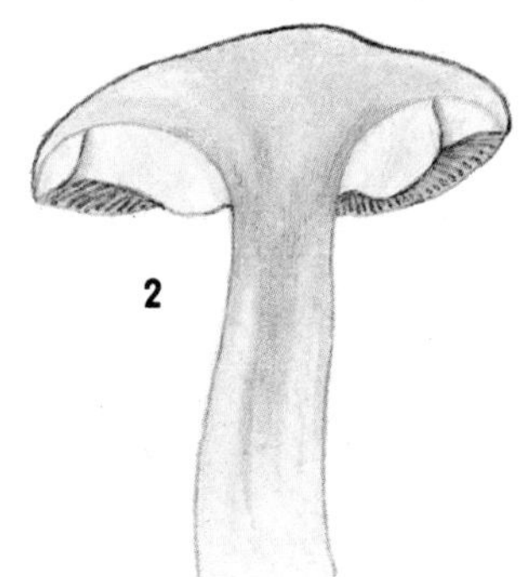


*Tricholompsis rutilans* (1) has cap 5—15 cm wide, stipe is 6—10/1—1.5 cm in size; whole fruit-body is yellow, velvety. Cap has often dark wine-red scales; gills golden yellow. Flesh yellowish (2); taste mild or bitterish, aroma sour to stale. Spore powder white; spores (3) colourless, 6—8/4.5—5.5 µm. Cystidia on the edges of gills are elongated and club-shaped (4).

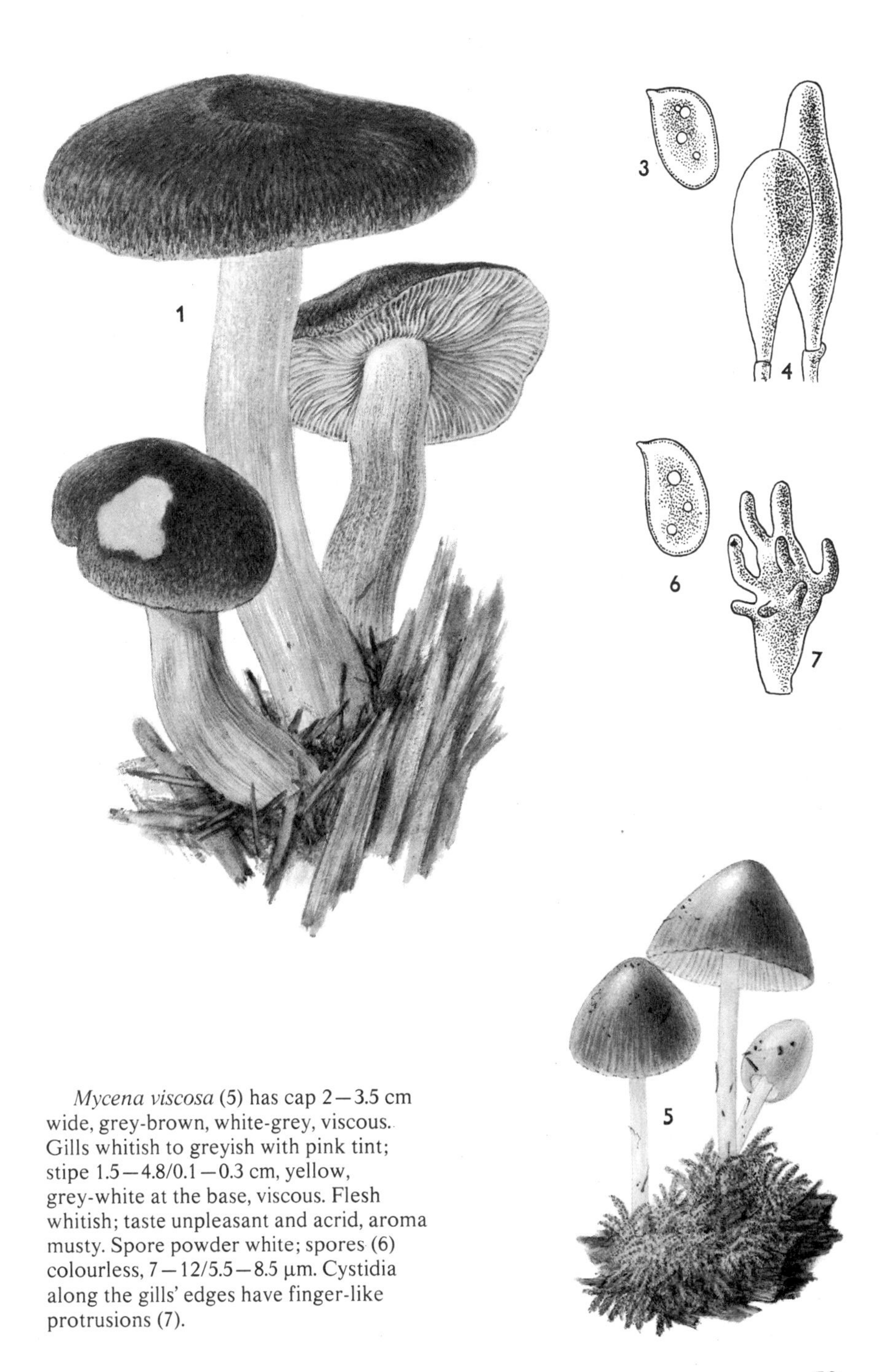


*Mycena viscosa* (5) has cap 2—3.5 cm wide, grey-brown, white-grey, viscous. Gills whitish to greyish with pink tint; stipe 1.5—4.8/0.1—0.3 cm, yellow, grey-white at the base, viscous. Flesh whitish; taste unpleasant and acrid, aroma musty. Spore powder white; spores (6) colourless, 7—12/5.5—8.5 µm. Cystidia along the gills' edges have finger-like protrusions (7).

## Soap Tricholoma *or* Strong-scented Agaric

*Tricholomataceae*

*Tricholoma saponaceum* (FR.) KUMM.

The genus *Tricholoma* is extremely rich in species; in Europe there are about eighty. The Soap Tricholoma is the most abundant. It inhabits all types of forests, growing under spruce, pine and fir trees. It is less frequent in mixed and deciduous woods. It can be found at high altitudes, even at 2,000 m above sea level. It has been located in almost every continent, but is not considered good eating because of its strong soapy smell. Since the presence of poisonous substances has not been excluded, it is not recommended for collection. The Soap Tricholoma has a variable shape and colouring. The colour of the cap ranges from light grey, dark grey, brown-grey to light olive green, yellow-green or even reddish and red-brown. The stipe is white at the top, becoming pale pink when old; it can be short and stout or long and thin. Reliable features for identifying this species are the soapy smell and the fact that the flesh and often the entire fruit-body turns pink or red.

The Soap Tricholoma slightly resembles *T. sejunctum,* a poisonous mushroom which grows in deciduous and coniferous forests and has a yellowish or greenish cap and brownish, dark striate centre. Its gills are white, or sometimes yellowish; the stipe is white. The flesh does not change colour when exposed to air; the aroma is not soapy but floury, and the taste is bitter.

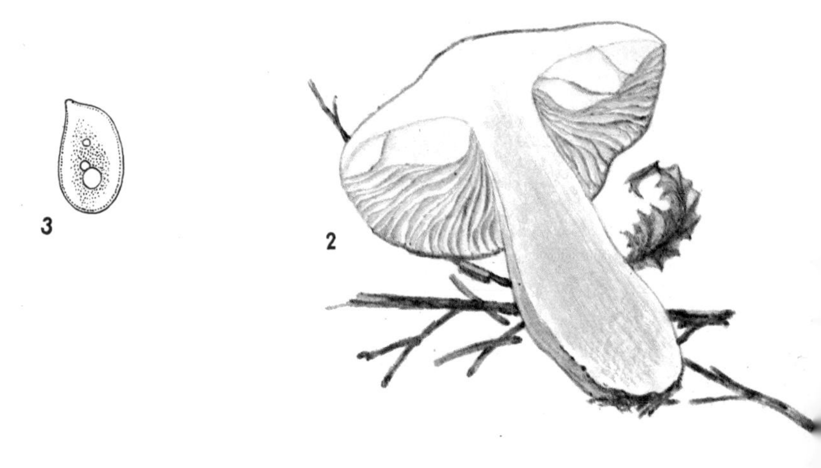

*Tricholoma saponaceum* (1) has cap 5—10 cm wide, fleshy, variable in colour, smooth or with fine scales and with deeply involuted margin. Gills pale, usually green-yellowish, yellowish-grey or waxy yellow, sinuate-decurrent (4) or attached with a curve. Stipe 5—10/1—3 cm, whitish or grey, often with the same shades as the cap, reddish, pruinose, later glabrous, but also with fine grey squamous fibrils, sometimes rooty, spindle-shaped. Flesh white or pale, often turning faint pink on the stipe (2); taste unpleasant, often bitterish, aroma always soapy and unpleasant. Spore powder white; spores (3) colourless, 5—6/3.5—4 µm.

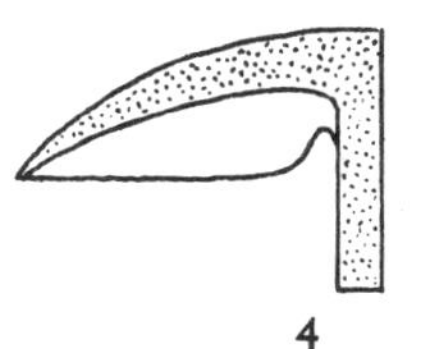

## Firwood Agaric

*Tricholomataceae*

*Tricholoma auratum* (PAUL. ex FR.) GILL.
Syn.: *T. equestre*

*Tricholoma auratum*, an excellent edible mushroom, is sold in country markets in some parts of Europe. It is characterized by its green-yellow or brown-yellow cap and stipe and white flesh, which has a pleasant, floury aroma. *T. auratum* grows from September to November over the entire temperate zone of the northern hemisphere, forming mycorrhizal associations with pines.

*T. flavovirens*, a smaller species with lemon gills and a thin stipe, grows in coniferous and deciduous forests. Both species congregate in small clusters, sometimes together with the Sulphur Tricholoma *(T. sulphureum)*, which is poisonous. The Sulphur Tricholoma can be recognized by its sulphur yellow cap and the cross-section through its stipe. It lives in mycorrhizal association with various oak species and also with beech. Apart from deciduous and mixed forests it can be also found in coniferous growths.

*Tricholoma auratum* can be confused with the rare *T. flavobruneum*, but this species can be identified by its finely squamous cap and reddish, fibrilous stipe. The fruit-body is brown-yellow or red-yellow with yellowish gills, and the taste is floury. It grows in mycorrhizal association with birch in deciduous and mixed forests from the beginning of summer to the end of autumn. It is edible, but its quality is poor.

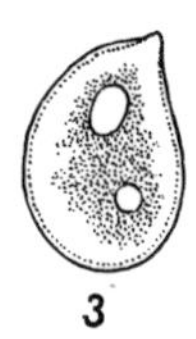


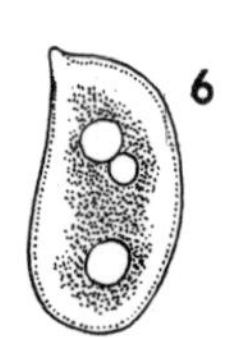


*Tricholoma auratum* (1) has cap 5—10 cm wide, green-yellow or brown-yellow, with deep dark fibrils. Gills sulphur yellow; stipe 4—10/1—3 cm, sulphur yellow or yellow-white. Flesh white, yellow along the edges of a cut (2); taste and aroma are floury. Spore powder white; spores (3) colourless, 6—7/3—4 μm.

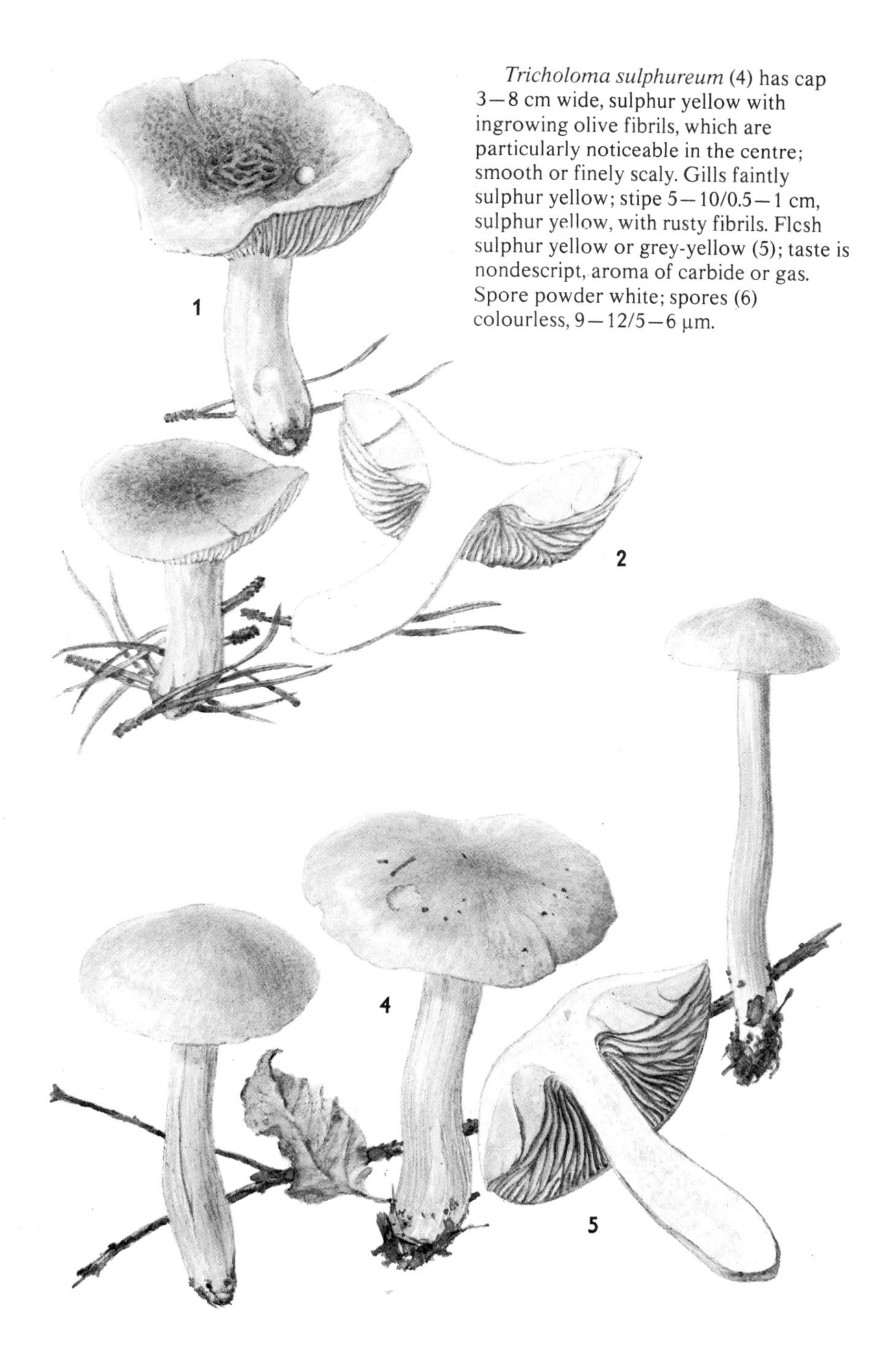


*Tricholoma sulphureum* (4) has cap 3–8 cm wide, sulphur yellow with ingrowing olive fibrils, which are particularly noticeable in the centre; smooth or finely scaly. Gills faintly sulphur yellow; stipe 5–10/0.5–1 cm, sulphur yellow, with rusty fibrils. Flcsh sulphur yellow or grey-yellow (5); taste is nondescript, aroma of carbide or gas. Spore powder white; spores (6) colourless, 9–12/5–6 μm.

## Dingy Agaric

*Tricholomataceae*

*Tricholoma portentosum* (FR.) QUÉL.

The Dingy Agaric is well-known and much collected for its excellent flavour. It grows in autumn in coniferous, pine and spruce forests in sandy soil, always in large numbers. It lives in mycorrhizal association with pine trees, throughout the temperate zone of the northern hemisphere.

The fragile grey fruit-bodies of *T. terreum* appear in autumn in forests, parks and gardens under coniferous trees, mostly pine, especially on lime-rich soil. It is the most common representative of the group of about 15 *Tricholoma* species with a grey or grey-brown cap. Some of the other members are *T. argyraceum*, with a light grey, scaly, later silky fibrous cap and yellow-patched gills; *T. scalpturatum*, with finely scaly cap and yellow gills and flesh; *T. pardinum* with a cracked, squamous cap and a stout stipe which turns brown-red when bruised; and *T. virgatum* with ingrown longitudinal fibrils on the cap and greyish flesh. All the above-mentioned species are edible with the exception of the poisonous *T. pardinum* and *T. virgatum*.

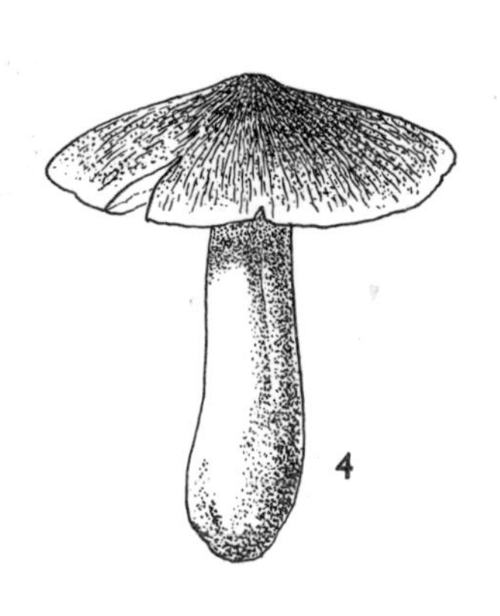


*Tricholoma portentosum* (1) has cap 4–10 cm wide, light grey to black-grey, often with yellowish, greyish or lilac tones and with blackish ingrown radiating fibrils. Gills (2) white with lemon yellow tinge; flesh whitish or greyish; taste and aroma slightly floury. Spores (3) colourless, 5–6/4–5 µm.

The poisonous *T. sejunctum* (4) is distinguished by its yellow-green to yellowish cap, which has a brownish centre and a white stipe. Taste floury at first, later becoming rather bitter. It grows in mixed forests, most often under oak trees.

*Tricholoma terreum* (5) has cap 3–8 cm wide, grey-brown, porphyry grey, dark striate, slightly scaly along the edges. Gills are white-grey (6); stipe is 4–8/1–1.5 cm, whitish to greyish, fibrilous to scaly. Taste nondescript, aroma earthy. Spores colourless, 5–8/4–5 µm.

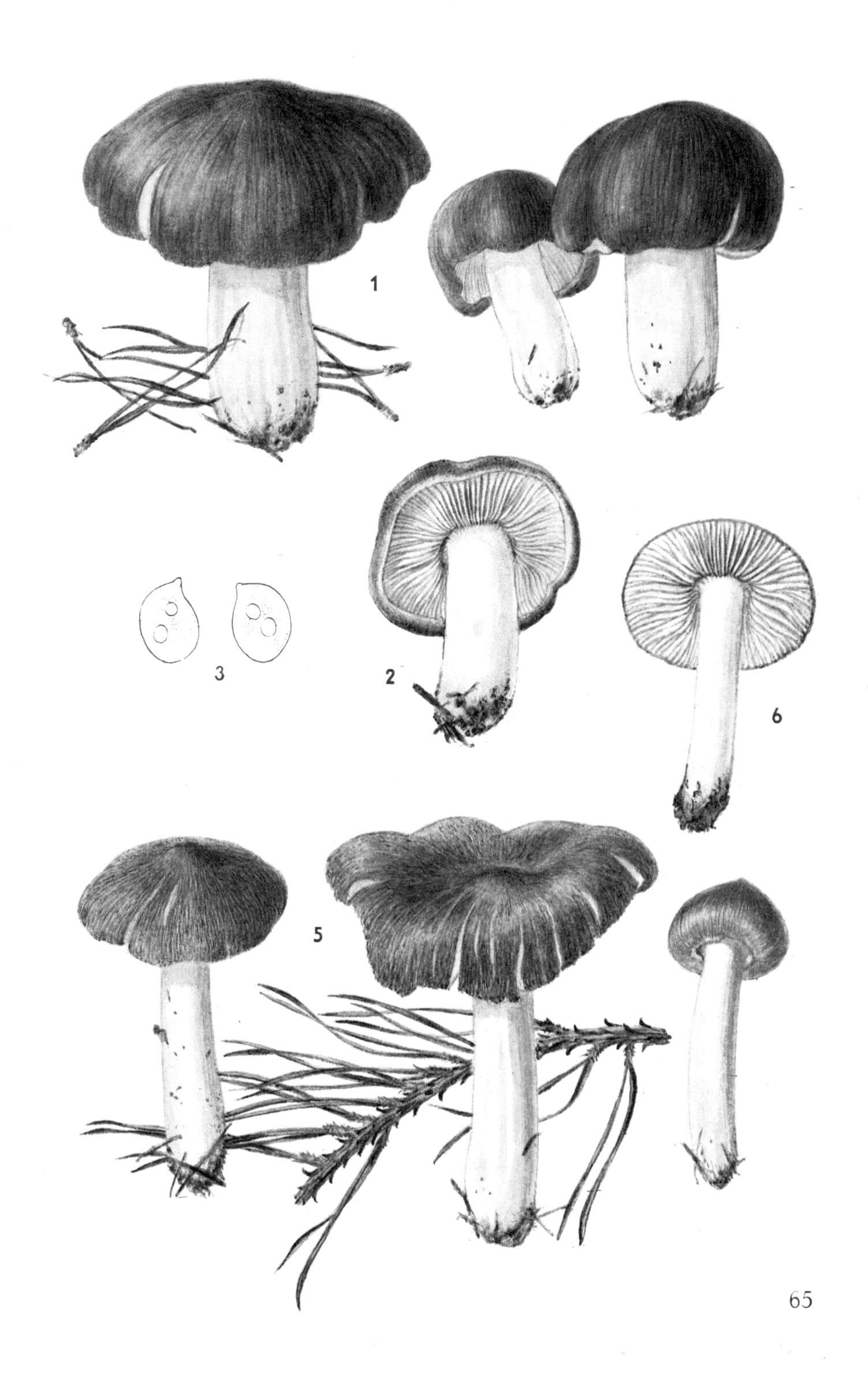
1
3
2
6
5

## Wood Agaric

*Tricholomataceae*

*Collybia dryophila* (BULL. ex FR.) KUMM.

*Collybia dryophila* is a much variable species. It remotely resembles the Fairy-ring Champignon with which it is often confused, but it does not approach the latter's flavour. In the United States it is considered slightly poisonous for it sometimes causes stomach upsets. In Europe it is classified as an edible mushroom of lower quality. *Collybia dryophila* differs from the Fairy-ring Champignon by its hollow stipe, which is brownish to rusty at the base, whereas the Fairy-ring Champignon has an entire, pale ochre stipe.

*C. dryophila* is abundant throughout the temperate zone of the northern hemisphere. It thrives as a typical humus saprophyte in all types of forests, but most often in oak woods, from spring to autumn, and is especially abundant after rain. Fallen leaves are often completely covered with the white mycelium and in this way *C. dryophila* aids decomposition of the top soil. In dry weather the fruit-bodies dry out, but in damp conditions they partly regain their original shape.

A similar species is *C. exsculpta,* which has a red-brown or chocolate brown, hygrophane cap and lemon or sulphur yellow gills and stipe. It grows in coniferous forests and is edible.

*Marasmius wynnei* also resembles *C. dryophila;* however, it is distinguished by its characteristic greyish scalloped cap, grey or grey-purple gills and grey stipe with a black-brown base. This exclusively European species can be found in large numbers in summer and in autumn growing in the foliage and needles of all types of forests or even in grass outside the forest. It is edible.

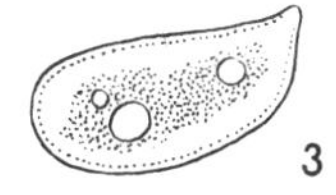


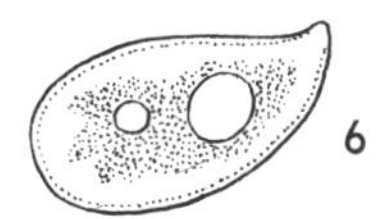


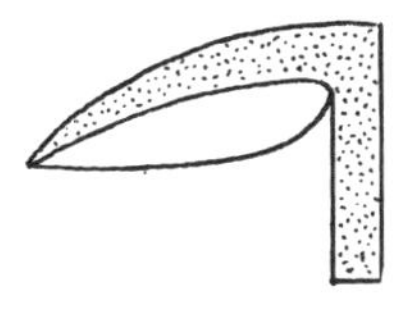


*Collybia dryophila* (1) has cap 2—5 cm wide, hygrophane, yellow-brown, light rusty with darker centre, pale ochre when dry. Crowded gills (7) adnexed, whitish, dingy, yellowish. Stipe 4—8/0.2—0.5 cm, ochre or brownish, often with rusty base, tubular. Flesh whitish (2); taste and aroma are slightly mushroom. Spore powder white; spores (3) colourless, 5—6/2—3 μm.

*Marasmius wynnei* (4) has cap 2—5 cm wide, hygrophane, at first white, later whitish, greyish and also purple-grey, with scalloped edges when mature. Gills adnexed, whitish, greyish or grey-purple (5,7); stipe 4—7/0.1—0.3 cm, black-brown at the base. Flesh whitish, taste and aroma indistinct. Spore powder white; spores (6) colourless, 6—7/3—3.5 μm.

## *Strobilurus esculentus*

(WULF. ex FR.) SING.

Syn.: *Collybia esculenta*



The small number of fungus species which grow on the cones of coniferous trees is represented by the genus *Strobilurus,* which includes only four spring species. *S. esculentus* can be found on old, partly buried spruce cones. Several fruit-bodies grow out of one cone; some of them are equipped with a shortly elongated, pubescent stipe. Pine cones which are buried in the top soil host *S. stephanocystis.* One or two fruit-bodies with a pubescent elongated underground stipe usually grow on one cone. In sandy pine forests, where cones are buried deep in the ground, the underground stipes are as much as 30 cm long. Both *Strobilurus* species are edible. *S. tenacellus* is another spring European species growing on pine cones. It is distinguished from *S. stephanocystis* by its bitter taste.

In North America the genus is represented by *S. conigenoides,* which appears on the cones of the Weymouth Pine *(Pinus strobus)* and spruce, and sometimes even on magnolia inflorescences. This species has a yellow-brown cap, only 1—5 mm wide.

In summer and autumn spruce and pine cones bear *Baeospora myosura* with a cap 1—2 cm wide, which is flesh brown and scattered with white flakes. A rare, edible species, which begins to appear in spring, is *Mycena strobilicola;* it has grey-brown fragile fruit-bodies, club-shaped cap and a nitrate aroma.

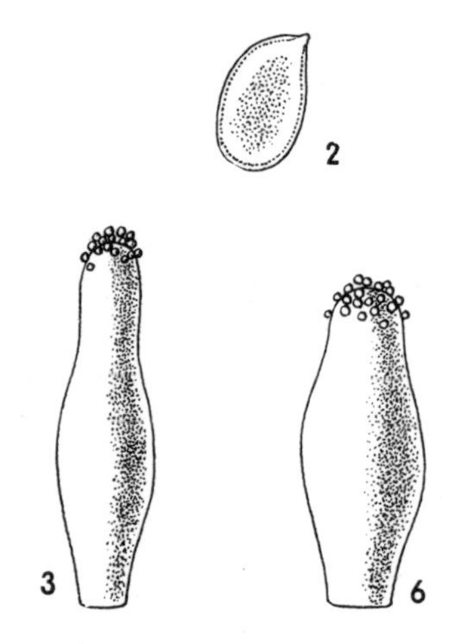


*Strobilurus esculentus* (1) has cap 0.8—2 cm wide, light ochre, almost chocolate brown in damp weather, smooth, with whitish gills. Stipe 1—3/0.1—0.3 cm, rooting, ochre brown to yellow-brown; flesh white. Spores (2) colourless, 5.3—7/3—4 μm. Cystidia have minute granules at the apex (3).

*Strobilurus stephanocystis* (4) has cap 0.8 – 1.8 cm wide, ochre or dark brown. Gills white to creamy ochre. Stipe is 1 – 4/0.1 – 0.3 cm, deep rooting, yellow-ochre to red-ochre, white on the lower surface of the cap; flesh white (5). Spores resemble those of *S. esculentus* but are 6 – 8/3 – 4 μm in size. Cystidia have coarse granules at the apex (6).

Both species have insipid taste and smell. Spore powder is white.

## **Porcelain Fungus** *Tricholomataceae*

*Oudemansiella mucida* (SCHRAD. ex FR.) HOEHN.
Syn.: *Collybia mucida*

The genus *Oudemansiella* has about 10 species in Europe, all edible, but generally very rare. Only *O. mucida, O. radicata* and *O. longipes* are more common. All *Oudemansiella* species belong to wood fungi widespread in the entire temperate zone of the northern hemisphere.

The Porcelain Fungus appears most often in autumn, in clusters or individually, on living or dead trunks and branches of beech trees, and less frequently on oak. It is typical of mountainous beech growths. Its fruit-bodies, glistening under the rays of the autumn sun high on the tree trunks, seem as if made of porcelain. Their white colour later becomes grey-brown. In laboratory conditions the mycelium produces an antibiotic substance mucidin with a strong antifungal reaction, which is currently produced in many countries and is effectively used in the treatment of ailments caused by fungi.

The fruit-bodies of the Rooting Shank *(O. radicata)* grow individually in summer and autumn on stumps and roots of deciduous trees in forests, parks and gardens. They have a characteristic spindle-shaped stipe (pseudorhiza), which can be as much as 30 cm long depending on the depth of the root which hosts the mushroom. The same conditions and season suit the closely related *O. longipes*. Slightly bitter in taste, it has a rusty, grey-brown, dry cap and a velvety stipe.

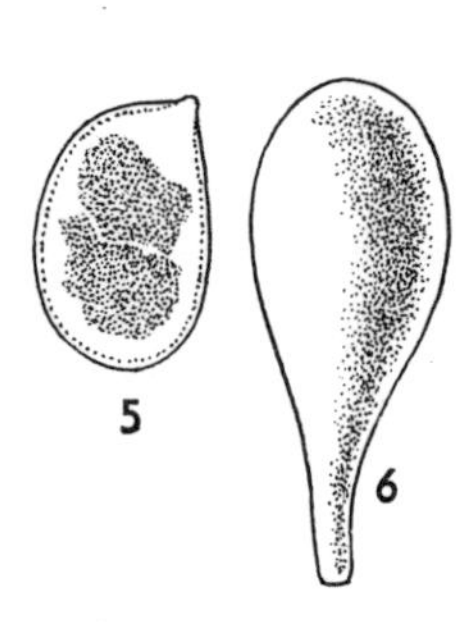


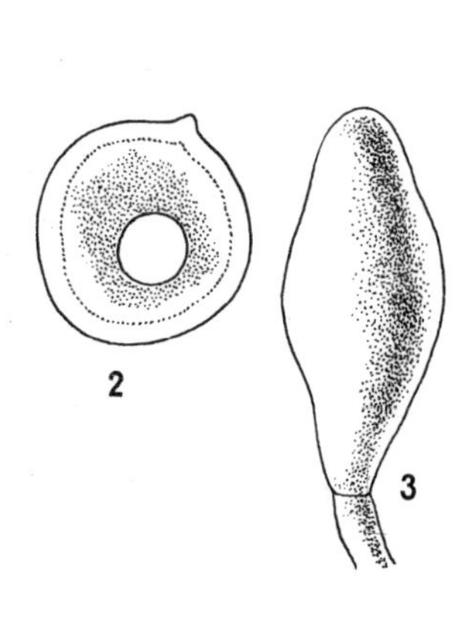


*Oudemansiella mucida* (1) has cap 3–8 cm wide, slimy, smooth, later wrinkled, grooved, white, brownish when old. Gills white; stipe 4–9/0.3–0.8 cm in size, slimy, white, with a white ring. Spore powder white; spores (2) colourless, 14–18 μm. Cystidia broadly spindle-shaped (3).

*Oudemansiella radicata* (4) has cap 2.5—10 cm wide, slimy, wrinkled, whitish, ochre to brown. Gills white; stipe 8—20/0.5—1 cm, bare, deeply rooting, whitish, brownish at base. Spore powder white; spores (5) colourless, 12—15/9—11 µm. Cystidia are club-shaped (6). Flesh of both species is white, flavour and aroma are nondescript. Gills of all *Oudemansiella* species are adnexed (7).

## **Broad-gilled Tricholoma** *Tricholomataceae*

*Megacollybia platyphylla* (PERS. ex FR.) KOTL. et POUZ.
Syn.: *Collybia platyphylla*

The Broad-gilled Tricholoma is a wood and saprophytic mushroom, widespread throughout the temperate zone of the northern hemisphere. In North America it is known under the name *Tricholomopsis platyphylla.* The fruit-bodies appear in summer and autumn individually or in small clusters on damp, decayed stumps or dead trunks of beech, birch and alder trees, which are overgrown with moss and lichens. Less frequently, it is found on coniferous trees. Most often it inhabits mountainous beech forests; it is rare in lowlands. The Broad-gilled Tricholoma is characterized by its grey-brown cap, which can be as much as 20 cm wide and has its surface covered with prominent radiating filaments, and by its broad, well-spaced gills. The base of the stipe produces a cord-like mycelium (rhizoids), which is white at first, becoming grey-black later. The mycelium is several tens of cm long, and penetrates rotten wood. The Broad-gilled Tricholoma is edible but can taste bitter.

The slightly poisonous *Collybia fusipes* lives only in Europe and northern Africa. It grows in summer on roots of dead oak trees and their stumps. The fruit-bodies always appear in clumps, as the lower thin sections of their stipes grow together. *C. fusipes* is a varied species, but it can be easily recognized by its red-brown colouring and long, spindle-shaped stipe.

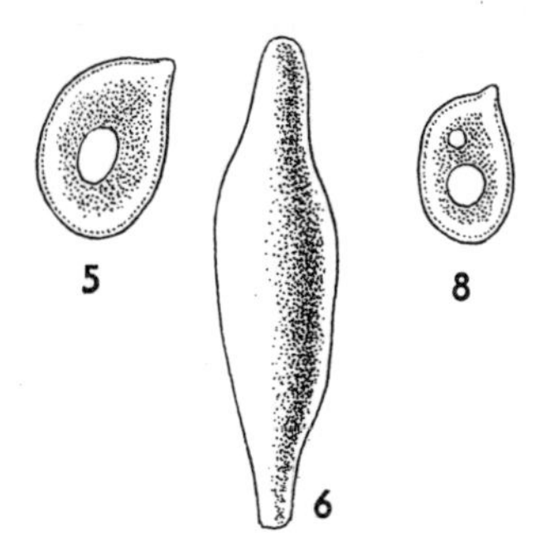


*Megacollybia platyphylla* (1) has cap usually 6—12 cm wide, grey-brown, with ingrown, radially streaked fibrils. Adnexed gills (3) 15—25 mm wide, whitish or yellowish, with rough, irregularly serrated edges, sometimes brown (2). Stipe 5—15/0.9—1.5 cm, dirty white to grey with dark fibrils, base as if cut off, with rhizoids (4). Flesh white (2); taste mild and aroma nondescript. Spore powder white; spores (5) colourless, 7—8/6—7 mm. Cystidia bottle-shaped (6).

*Collybia fusipes* (7) has cap 4 – 7/10 cm wide, dark red-brown. Stipe is 8 – 12/1 – 2 cm, with spindle-shaped rooting system, dark red-brown. Gills white-grey or reddish, often with reddish patches; flesh pale, taste and aroma nondescript. Spore powder white; spores (8) colourless, 4 – 5.5/3 – 3.5 μm.

## Winter Fungus, Velvet Shank

*Tricholomataceae*

*Flammulina velutipes* (CURT. ex FR.) SING.
Syn.: *Collybia velutipes*

The Winter Fungus is a true winter species, growing almost worldwide from November to March. The fruit-bodies appear as soon as the temperature drops to freezing point. During fierce frosts the fruit-bodies do not increase in volume, but frozen solid and sometimes covered with snow, they are very well preserved. When the fruit-bodies defrost, they do not decay but continue to grow, and in fact can withstand several frosts and thaws. The Winter Fungus is exclusively a wood-species which grows on dead or living trunks and branches of deciduous trees (willow, poplar and lime) and exceptionally also on conifers. The fruit-bodies often form large clusters of individuals at different stages of development, ranging from small subglobose caps to expanded ones which release white spore powder that settles on slimy neighbouring caps.

An excellent edible mushroom, the Winter Fungus is cultivated on cut wheat straw or in the open air on sawn logs or stumps of lime and willow trees, which are watered with a mixture of pulped caps. In Japan a substance called flammulin has been isolated from the Winter Fungus. This has been shown to retard cancerous tumours in mice and may prove valuable in the treatment of humans.

The Winter Fungus is closely related to *F. ononidis,* which grows individually on underground sections of *Ononis spinosa.*

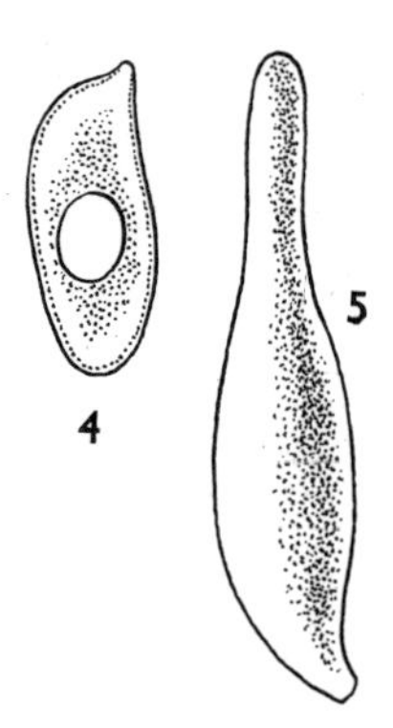


*Flammulina velutipes* (1) has cap 2—8 cm wide, yellow-brown, yellow-ochre, chestnut, slimy. Gills rounded-adnexed (2), white or creamy, stipe 3—10/0.3—0.8 cm, tomentose, dirty white or creamy at the top, rusty yellow-brown in the middle and chestnut brown or rusty brown at the base (3). Flesh creamy, taste and aroma nondescript. Spore powder white; spores (4) colourless, 6.7—10.3/2.5—4.5 μm. Cystida are spindle-shaped (5).

*Flammulina ononidis* (6) has cap 1.1—2.1 cm wide, stipe is 1.3—5/0.1—0.3 cm. Spores (4) 9.1—12.5/4—6 μm. Other features similar to those of the Winter Fungus.

1
3
6
2

## Fairy-ring Champignon

*Tricholomataceae*

*Marasmius oreades* (BOLT ex FR.) FR.

Some *Marasmius* and *Hygrocybe* species are typical grass mushrooms. The Fairy-ring Champignon has a flexible stipe and cap and pale ochre colouring. Its fruit-bodies appear in clusters, strips and circles in meadows and pastures, especially after warm summer rains. These 'fairy-rings' are several metres in diameter and increase in size every year as the mycelium spreads through the ground. As long as conditions remain the same, the Fairy-ring Champignon grows in one place for decades. It prefers manured, grassy spots, pastures, meadows and lawns in the neighbourhood of villages, also growing on field borders, forest edges and in spruce thickets. The Fairy-ring Champignon can be found from spring to autumn in Europe, America, Asia and Africa. The fruit-bodies last for a long time, as they dry out during drought and regain their original shape in damp weather. They are edible and have an excellent flavour.

The Fairy-ring Champignon is related to the heat-loving *M. collinus,* which also grows outside forests, but is rarer. It can be recognized by its hollow stipe.

The fruit-bodies of the Conical Wax-cap *(Hygrocybe conica)* add colour to lawns and meadows, when they appear after heavy rainfall in autumn. However, their gay colouring soon changes into dark purple, almost an ink black colour. Another species which turns black is *H. nigrescens,* which has a larger, blunt club-shaped, fibrilous, orange red cap with a lemon or orange yellow stipe. Both species are slightly poisonous.

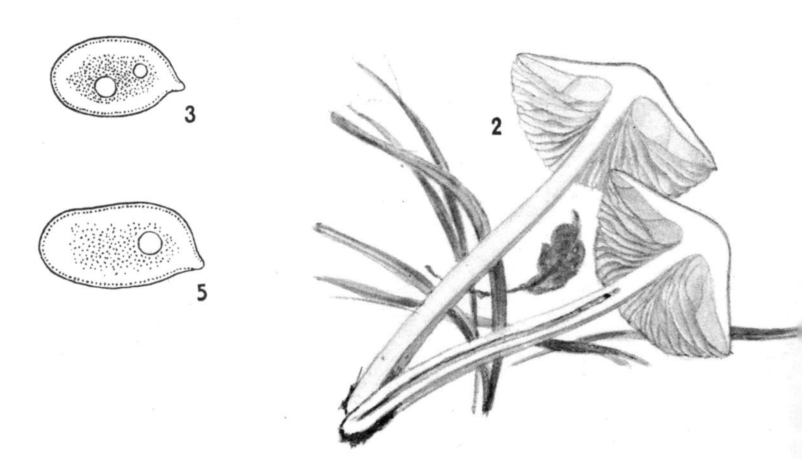

*Marasmius oreades* (1) has cap 2—5 cm wide, hygrophane, beige, pale ochre, red-brown, often with striate margin. Gills well-spaced, creamy, ochre in damp conditions. Stipe 4—7/0.2—0.4 cm, full, whitish, pale, pruinose. Flesh whitish (2); taste and smell pleasantly fungoid. Spore powder white; spores (3) colourless, 7—9/4—6 μm.

*Hygrocybe conica* (4) has cap 2—6 cm wide, glabrous, slimy, sharply conical, yellowish-red, orange or scarlet red. Gills yellowish; stipe 5—10/0.4—0.9 cm, fragile, yellow, orange, ink black at the base. Entire fruit-body becomes black when bruised or when old. Flesh (6) yellow, later turning black; taste and smell nondescript. Spore powder white; spores (5) colourless, 8—13/5—8 μm.

## Garlic Marasmius

*Tricholomataceae*

*Marasmius scorodonius* (FR.) FR.

The fruit-bodies of *M. scorodonius* measure only several centimetres and appear only after heavy rainfall in summer and autumn. Their distribution is almost worldwide. Hundreds of small fruit-bodies grow out of the top soil of spruce and pine needles and branches in coniferous forests and heaths. *M. scorodonius* has a garlic taste and smell which grow stronger if the fruit-body is crushed. It is an edible mushroom, often added to sauces and soups.

A similar species, *Micromphale perforans,* is found especially in mountainous coniferous forests, where it grows on needles of spruce and pine trees. The needles remain attached to the base of the stipe when the fruit-body is pulled out of the ground. These mushrooms appear in damp weather after rain throughout the year, even in winter if the weather is mild. Characteristic features, which distinguish it from *M. scorodonius,* include a pubescent stipe (the hairs are visible only when magnified) and fleshy gills. The unpleasant smell of *M. perforans* makes it inedible.

The Bonnet Mycena *(Mycena galericulata)* is abundant in lowlands and highlands of the whole temperate zone of the northern hemisphere, where it occurs on stumps and wood mostly of deciduous trees (beeches, oaks and alders). It fructifies throughout the year. It is edible, with a characteristic cucumber floury taste when fresh.

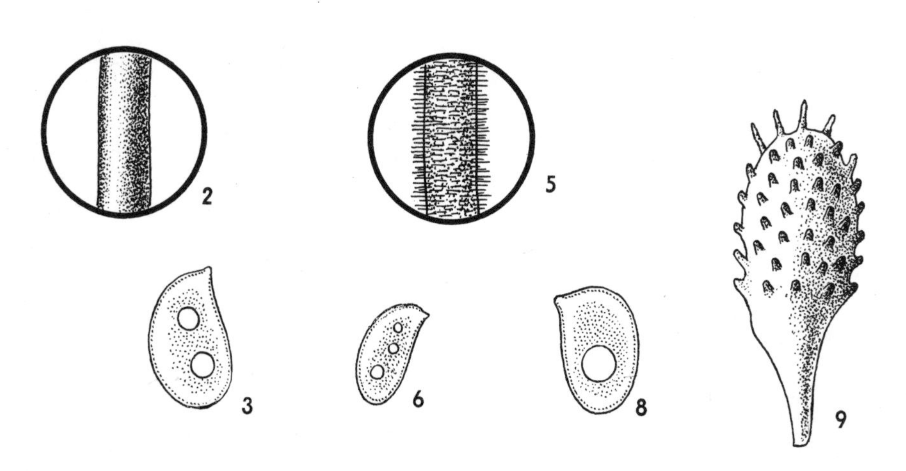

*Marasmius scorodonius* (1) has cap 1—2.5 cm wide, smooth, red-brown. Gills white; stipe 2—5/0.1—0.2 cm, smooth (2), red-brown. Taste and smell resemble garlic. Spore powder white; spores (3) colourless, 8—9/4—5 µm.

*Micromphale perforans* (4) has cap 0.5—1 cm wide, wrinkled, flesh brown. Gills light, fleshy; stipe 1—4/0.05—0.1 cm, ochre at the top, red-brown lower down, hairy (5). Aroma unpleasant. Spore powder white; spores (6) colourless, 5.5—6/3 µm.

*Mycena galericulata* (7) has cap 1—7 cm wide, hygrophane, light brown, grey-brown. Gills whitish, greyish, often turning pink when old. Stipe 3—15/0.2—0.8 cm, whitish, greyish, brownish at the base. Spore powder white; spores (8) colourless, 7.7—1.5/5.1—7.8 µm. Cystidia along gill edges have projections (9).

## Clean Mycena



*Mycena pura* (PERS. ex FR.) KUMM.

The Clean Mycena is the most widely distributed of all Gill fungi. This saprophytic mushroom grows in lowlands and highlands in all continents, even within the arctic circle in Greenland and Alaska. In the Alps and the Caucasus it has been found in low shrubberies 3,000 m above sea level. Its fruit-bodies appear in large clusters in summer and autumn in all types of forests with thick humus; in northern Africa it grows even in cedar woods. The Clean Mycena was considered edible in the past, but as a result of a large number of poisonings, it is now classified poisonous; it is slightly hallucinogenic as it contains psychotropic indole substances. Its colouring varies a great deal; fruit-bodies with a white cap belong to the variety *alba*, with a yellow cap to the variety *lutea* and with a greyish blue-green cap to the variety *multicolor*.

The Amethyst Laccaria *(Laccaria amethystina)*, with its dark violet gills, resembles the Clean Mycena. This edible mushroom has a flat cap with a central depression, but never cone-shaped or with an umbo, and it does not smell of radish. A closely related species, with radish aroma and pink colouring is *rosea*, which grows only in deciduous forests, most often on oak, beech and lime leaves. It is poisonous as it contains the alkaloid muscarine. Other related species with a radish aroma are either poisonous, as *M. pelianthina*, which is entirely purple-brown with serrated, purple-black edges; or not poisonous, such as the minute *M. pearsoniana* with purplish grey, arched decurrent gills.

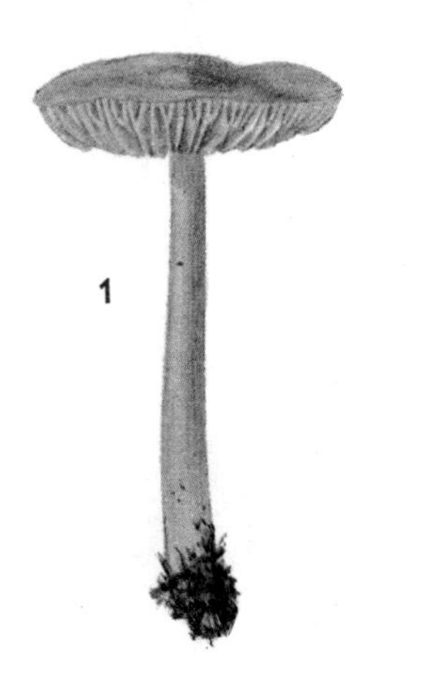


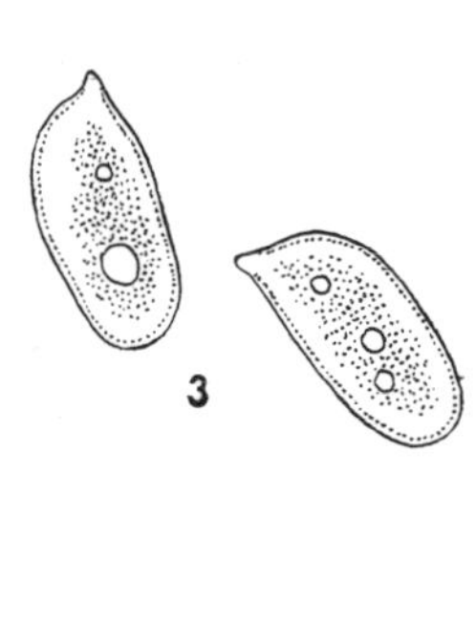


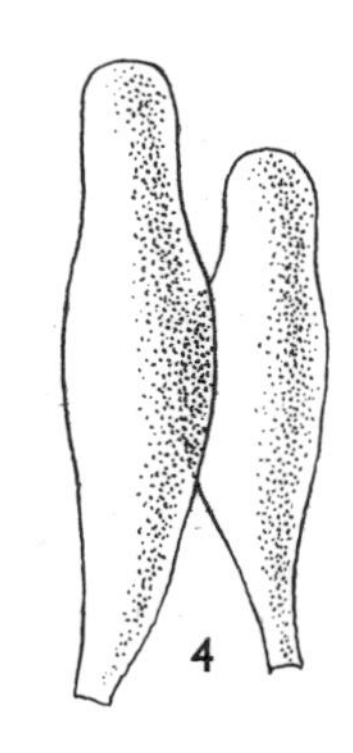

*Mycena pura* (1) has cap 1—4 cm wide, smooth, purple sometimes with bluish or yellowish tones. Gills adnexed or attached, with small denticle, white or purplish. Stipe 2—5/0.3—0.6 cm, purple.

*Mycena rosea* (2) has cap 2—5 cm wide, smooth, watery and transparent, pink. Gills white, sometimes pinkish; stipe is 2—6/0.3—0.7 cm, white, sometimes pinkish. Both species have white flesh, taste and aroma resemble radishes. Spore powder of both species white; spores (3) are colourless, 5—8.5/2.5—4 μm; cystidia are short and spindle-shaped (4).

## Blusher

*Amanitaceae*

*Amanita rubescens* (PERS. ex FR.) S. F. GRAY

The genus *Amanita* includes many poisonous, species, some even deadly poisonous, and several edible species which are poor in quality. Only the Blusher is ready good eating when cooked and has an excellent flavour. It should be gathered only by collectors who are familiar with all stages of its development; at the same time it is necessary to know all the poisonous *Amanita* species. Small fruit-bodies and their embryos are best avoided. Mycologists have classified several varieties of Blusher according to the size, colouring and arrangement of warts on the cap. These deviations are caused by habitat and weather, and are not permanent. The distribution of the Blusher is immense; it includes North America, northern Africa, Europe and Asia. It grows in abundance from June to autumn in all types of forests and soils with the exception of peat-bogs, living in mycorrhizal association with oak, beech, pine and spruce. The Blusher, according to some experts, is poisonous when raw; it contains phallolysin, a substance that causes the dissolution of red blood cells. But because phallolysin is thermolabile, well-cooked Blushers are harmless. The Blusher can be confused with the strongly poisonous Panther Cap, which has a brown-ochre or brown-olive cap, fine, glabrous ring and a stipe with a volva. The Panther Cap has no red-brown colouring on any part of its fruit-body.

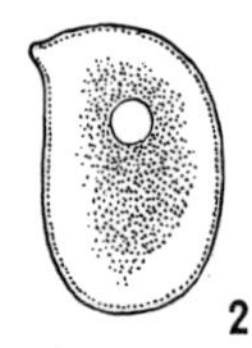


*Amanita rubescens* has cap 5—15 cm wide, light red-brown, brownish flesh red, with red-brown patches when old; margin smooth, surface covered with white warts. Gills free, swollen, white, later pinkish, often with red patches when old. Stipe 8—20/1—2.5 cm, white, light rusty red, tuberous; ring is large, white, pinkish, longitudinally striate. Volva missing, bulbous base encircled with reddish warts (1). Flesh white, reddish, sometimes with wine red patches, most often at the base of stipe; taste and aroma pleasantly fungoid. Spore powder white; spores (2) colourless, 8—9/6—7 µm.

1

## Stout Agaric

*Amanitaceae*

*Amanita spissa* (FR.) KUMM.

Though edible, *Amanita spissa* is of inferior quality. In summer and autumn it grows individually or forms clusters, either in coniferous forests where it forms mycorrhiza with pines and spruces, and also in deciduous woods, most often under oaks. Its common variety, *excelsa,* has stout grey fruit-bodies, a dirty white stipe, which is slightly swollen at the base and has a long spindly shape. In the variety *valida* the stipe and gills turn brown when bruised.

*A. spissa* is widespread throughout the temperate zone of the northern hemisphere and also in Australia. It is often confused with the poisonous Panther Cap or less frequently with the rarer *A. porphyria,* but may be distinguished by its almost smooth, grey-purple cap, disappearing, smooth, grey-purple ring, and by its stipe which swells at the base into a large soft bulb.

A related species, *A. aspera,* is also edible. Its characteristic features include a light or tan brown cap and a whitish or yellowish stipe, which has no bulb and which has yellowish lines or scales under its ring; the striate ring has a yellow margin. *A. aspera* mostly grows under beech trees.

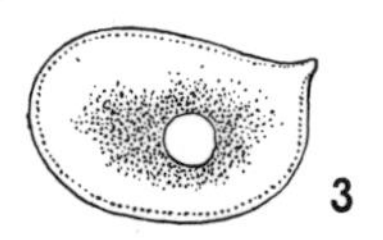


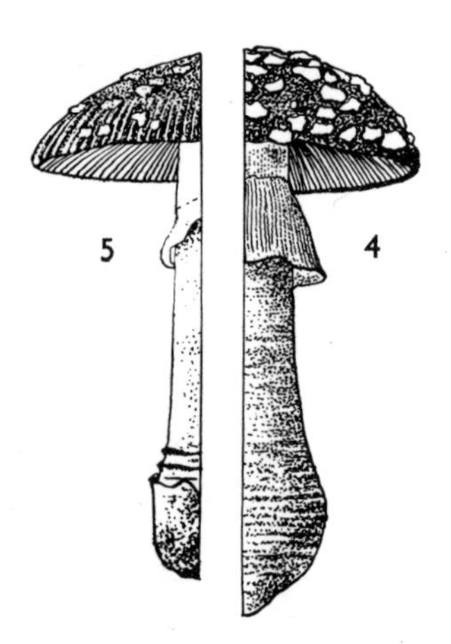


*Amanita spissa* has cap 7—12 cm wide, fleshy, greyish or brownish, with smooth margin; surface has firmly attached, white or greyish warts with concentric or irregular arrangement. Gills (1) white; stipe 8—12/1—2.5 cm, whitish, scaly, swollen at the base into an elongated bulb. Volva missing, but the bulb bears one or two incomplete circles of warts (2). Ring large, membranous, white, longitudinally striate. Flesh is whitish; taste nondescript, aroma of old potatoes. Spore powder white; spores (3) are colourless, 9—10/7—8 μm.

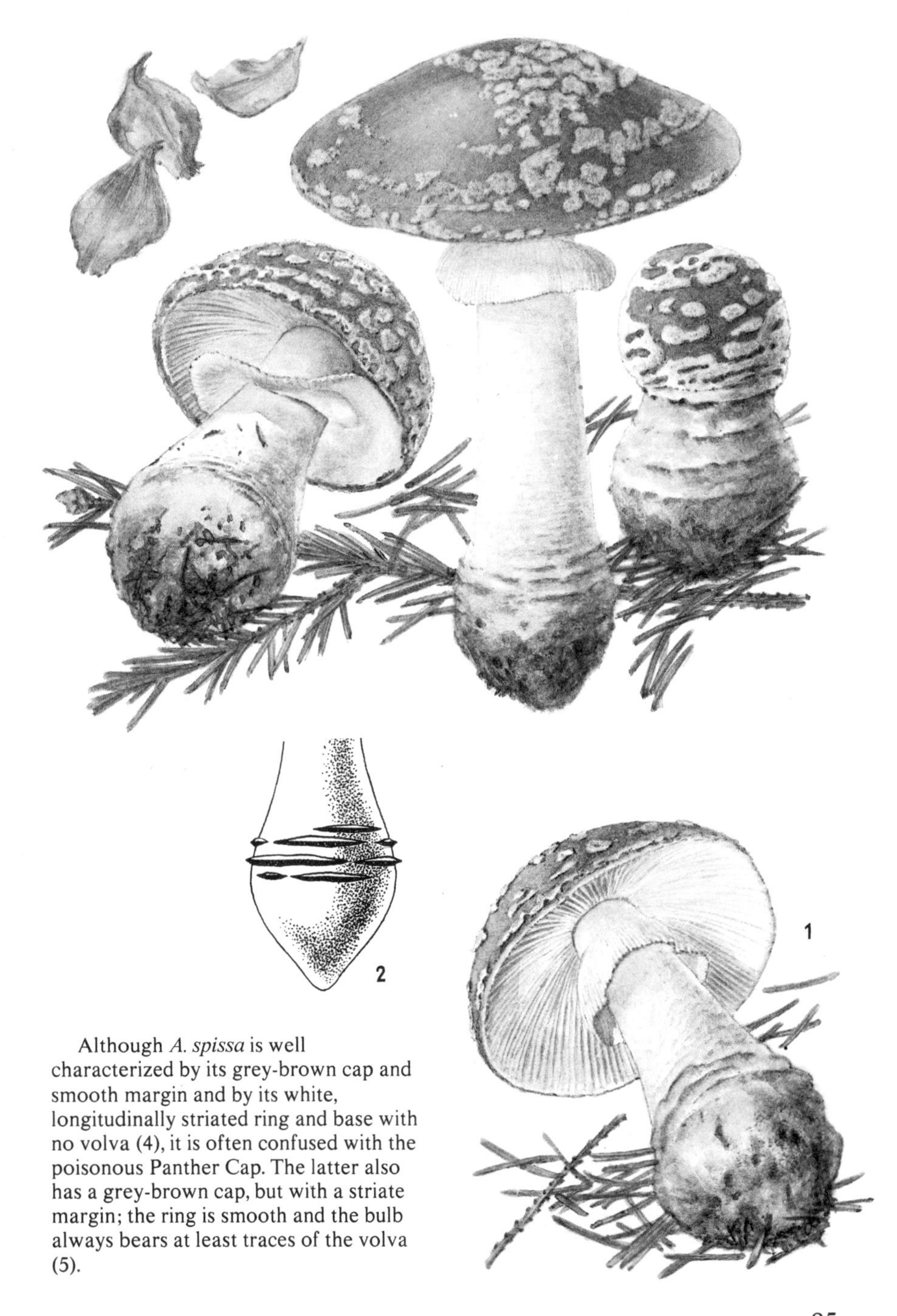


Although *A. spissa* is well characterized by its grey-brown cap and smooth margin and by its white, longitudinally striated ring and base with no volva (4), it is often confused with the poisonous Panther Cap. The latter also has a grey-brown cap, but with a striate margin; the ring is smooth and the bulb always bears at least traces of the volva (5).

# Death Cap



*Amanita phalloides* (FR.) LINK

The Death Cap and the Spring Amanita *(A. verna)* are the most poisonous mushrooms in the world. Both species contain several toxins, including phallin, phalloidin and the most toxic, amanitin. About 30 g of the fresh mushroom is a fatal quantity for an adult. Toxicity is not lessened by drying or cooking; only mushrooms pickled in vinegar are less toxic. Poisoning can also be caused by the spore powder.

The Death Cap can be confused with several species which have a green, greenish or bluish cap, such as *Russula aeruginea, R. virescens, R. cyanoxantha* or *Tricholoma auratum.* However, none of these has a stipe with a ring, a bulbous swelling at the base or a volva. The Spring Amanita can be confused with *Calocybe gambosa,* which is white, but its stipe lacks both a ring or a bulb with volva. The Spring Amanita and Death Cap are also frequently confused with white-coloured Field Mushrooms. However, the Field Mushroom has no volva or bulb, and its gills are pale only when the mushroom is very young, otherwise soon becoming pink and brown.

The Death Cap grows in summer and in autumn in deciduous forests, and also rarely in coniferous forests. It is most common in central Europe; it also grows in North and South America, Asia, Japan, China and New Zealand.

The Spring Amanita is a rare species. Its fruit-bodies can be found not only in spring, as can be by the common name of the species, but also grows from spring to autumn in deciduous forests with lime-rich soil.

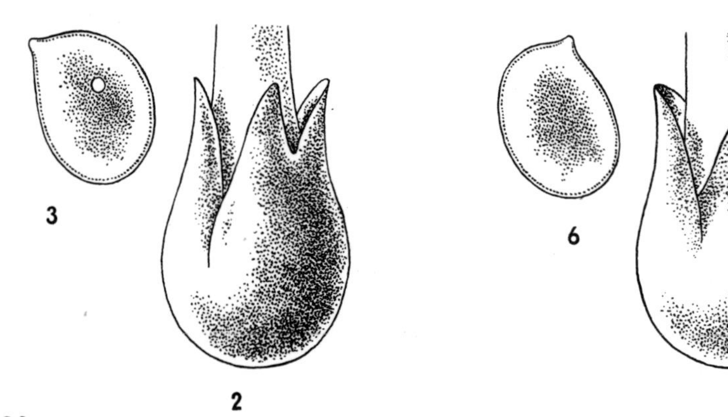

*Amanita phalloides* (1) has cap 5—14 cm wide, flat and expanded, sticky when moist, grey-green or yellow-green, streaked with yellow-brown radiating ingrown fibrils, very rarely with remnants of veil. Gills white; stipe 8—21/0.8—1.5 cm with whitish, yellowish or greenish striations, broadened at the base into a bulb with soft, whitish, irregularly lobed volva (2) which is buried in the ground; ring thin, white, smooth or with fine striations. Flesh white; slightly green under the cuticle; taste nondescript or faintly nutty, aroma sweet. Spore powder white; spores (3) colourless, 8—11/6.5—8.5 μm.

*Amanita verna* (4, 5, 6) resembles the Death Cap in nearly all features except for the colour of fruit-bodies, which are snow white; the surface of the cap lacks the ingrown fibrils.

## Destroying Angel



*Amanita virosa* (FR.) BERT.

The Destroying Angel equals the Death Cap and the Spring Amanita in toxicity. However, cases of poisoning by this species are less common as the Destroying Angel is quite rare. It appears in small groups in submountainous and mountainous areas in deciduous forests with acid soil, under beech, oak and birch trees. Exceptionally it grows under spruce. It has a spindle-shaped cap with flake-like remnants of veil on its margin and a woolly, scaly stipe which rarely has a ring. An important microscopic feature is the globular spores.

Only White Amanita is edible, but it cannot be recommended for collection as it is easily confused with other poisonous species of the genus. This is *Amanita alba,* which in all features except for its colour resembles the Common Grisette. *A. alba* has a stem which is swollen at the base, has no bulb or ring, and is white and fragile. The smooth white cap is always striated along the margin.

The Handsome Volvaria *(Volvariella speciosa)* at first sight resembles the white, poisonous *Amanita* species; however, its smooth stipe has no ring and the adult gills are not white but deep pink. It is an edible species which grows mainly outside forests, in humus and well manured soil in fields, gardens and and on old dumps and compost heaps.

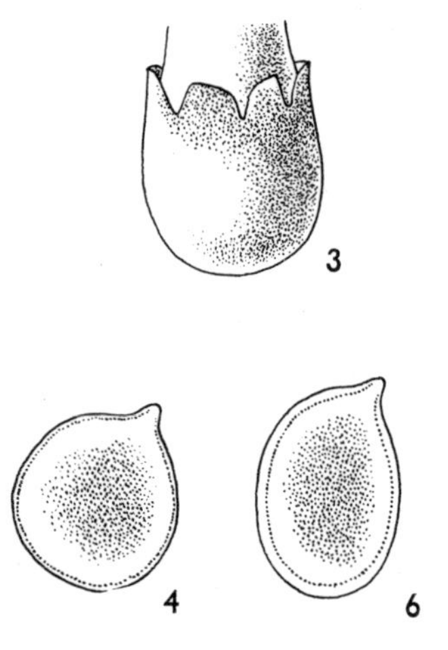


*Amanita virosa* (1) has cap 4—5 cm wide, sharply conical, white, sometimes with a brownish centre; surface has silky fibrils, sometimes with warts and always with remnants of veil on its margin. Gills white, free, swollen (2); stipe 8.5—15/0.6—1.6 cm, white, with disappearing ring, broadened at the base into a bulb. Surface of stipe from the ring downwards has white woolly scales; volva (3) broad, semi-free, silky, white. Flesh white, potassium hydroxide colours it orange; taste and aroma unpleasant. Spore powder white; spores (4) colourless, 8—11 μm.

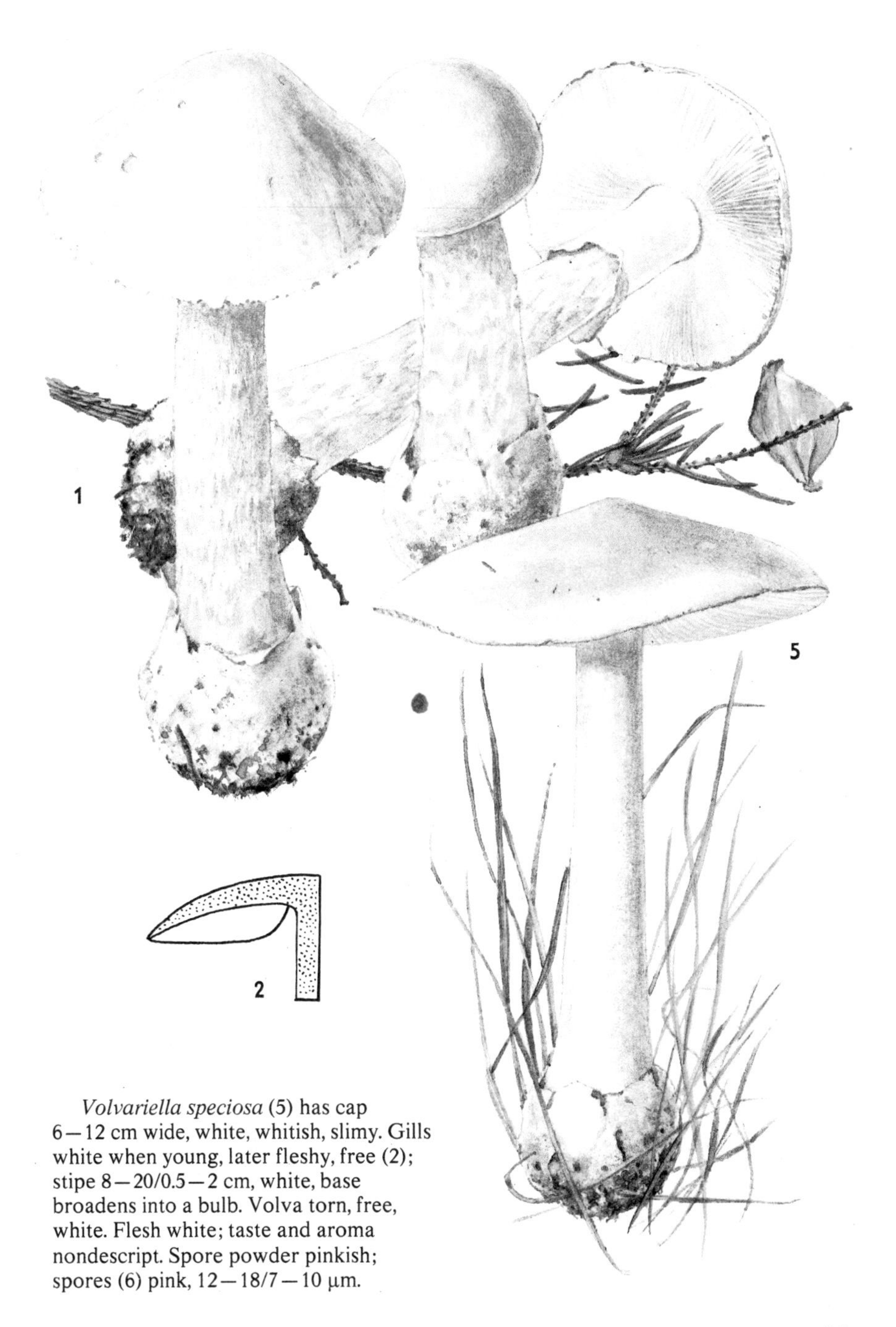


*Volvariella speciosa* (5) has cap 6—12 cm wide, white, whitish, slimy. Gills white when young, later fleshy, free (2); stipe 8—20/0.5—2 cm, white, base broadens into a bulb. Volva torn, free, white. Flesh white; taste and aroma nondescript. Spore powder pinkish; spores (6) pink, 12—18/7—10 μm.

## False Death Cap
*Amanita citrina* SCHFF. ex S. F. GRAY

Experts still argue whether *Amanita citrina* and *A. gemmata* are edible or poisonous species. It seems that poisoning does not occur if a small quantity is consumed. The False Death Cap probably contains a poisonous substance, bufotenin, and *A. gemmata* has a very low content of the toxins iboten acid and muscimol.

The False Death Cap can be seen from July to November in deciduous and coniferous forests of the lowlands and uplands. It is a common species which thrives in acid soil under oak, beech, birch, spruce and pine trees and forms a mycorrhizal association with them. It is most abundant in Europe and can be found also in North America, Asia and in northern and southern Africa. The False Death Cap has a smooth lemon yellow or green-yellow cap and the base of the stipe is enlarged into a loosely attached bulb with a volva. It closely resembles the deadly poisonous Death Cap, which has a green or green-yellow cap with yellow-brown ingrown fibrils, and a swollen stipe with a lobed volva.

*Amanita gemmata* grows in coniferous mixed forests from May to October. It is most abundant in sandy pine forests. This species prefers the warm climate of the temperate zone of the northern hemisphere. It is absent in the north and in mountainous and submountainous areas.

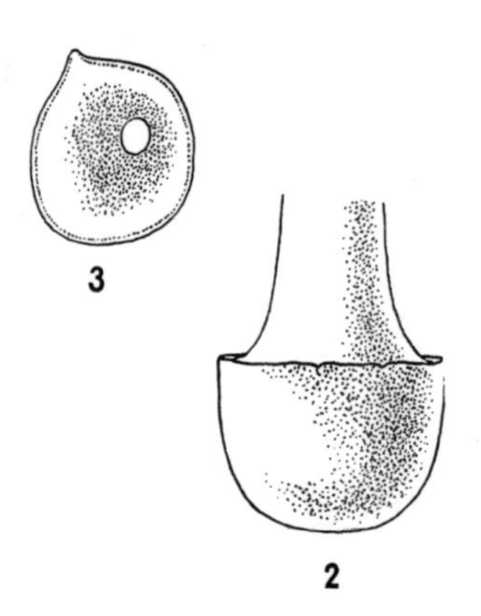


*Amanita citrina* (1) has cap 3—8 cm wide, lemon yellow, yellow-green, with pale yellow-brown warts; gills whitish with yellow tint. Stipe 5—12/0.8—1.5 cm, white or yellowish, with finely grooved ring; base of stipe enlarged into a loose bulb, without volva (2). Flesh white; taste unpleasant, aroma of raw potatoes. Spore powder white; spores (3) colourless, 7—10 µm.

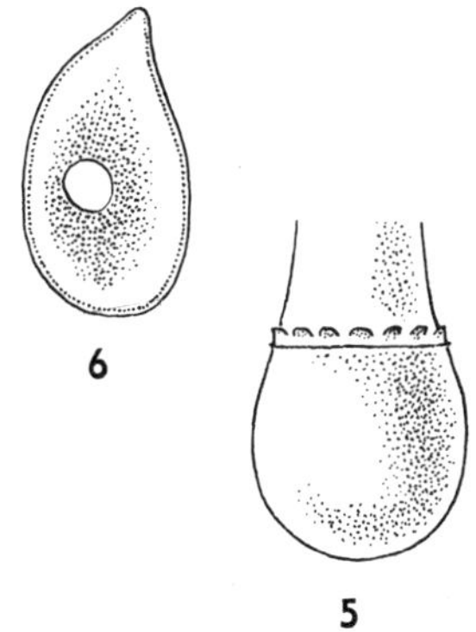


*Amanita gemmata* (4) has cap 5–11 cm wide, yellow, yellow-ochre; surface bare or has a sparse scattering of white warts, margin grooved. Gills white; stipe 6–15/1–2.5 cm, white, slightly enlarged at the base. Volva develops as a minute membranous circle (5). Ring forms membranous, cobweb-like remnants. Flesh white; taste and aroma are inconspicuous. Spore powder white; spores (6) are colourless, 10–12/7–8 μm.

## Fly Agaric

*Amanitaceae*

*Amanita muscaria* (L. ex FR.) HOOK.

The Fly Agaric is an intoxicating and poisonous mushroom whose effects were pointed out by travellers as early as the 17th century. They noticed that the men of some Siberian tribes either ate these mushrooms when dried or drank potion preparėd by boiling the caps and berries of bog whortleberry *(Vaccinium uliginosum)* in water or milk. Symptoms of intoxication appeared in about 30 minutes and were later followed by pleasant visual hallucinations and daydreaming, lasting about an hour. The active constituents, the alkaloid muscarine, iboten acid and muscimol, are passed into the urine or mother's milk without any change. Therefore where supplies of Fly Agaric were scarce, tribesmen drank their own urine for repeat performances.

The Fly Agaric grows in abundance in the entire temperate zone of the northern hemisphere. It can be found in summer and autumn in all forests, particularly in coniferous woods. The closely related *A. regalis* grows in submountainous and mountainous spruce forests, most often in northern Europe. It does not occur in southern Europe. It has the same toxic properties as the Fly Agaric.

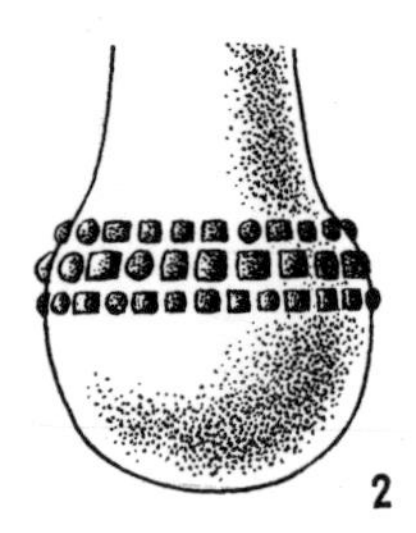


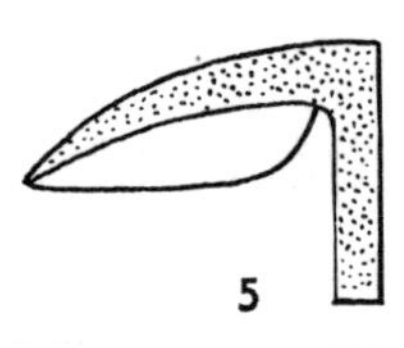


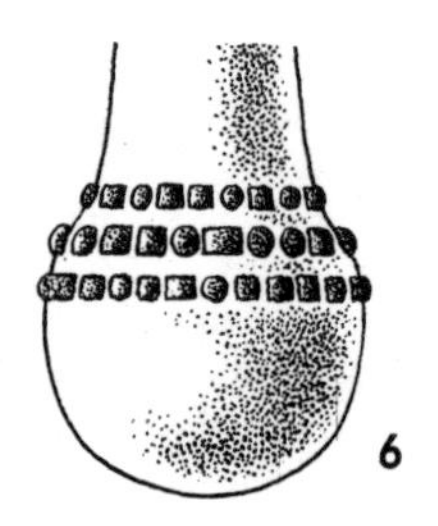


*Amanita muscaria* (1) has cap 5–15 cm wide, scarlet red or orange, with whitish warts. Gills white; stipe 7–20/1–2.5 cm, whitish; ring and bulb also whitish; bulb has 2 or 3 circles of warts on its surface (2). Flesh white, yellow under the cuticle. Spores (3) colourless, 8.5–12/7.5–9 µm.

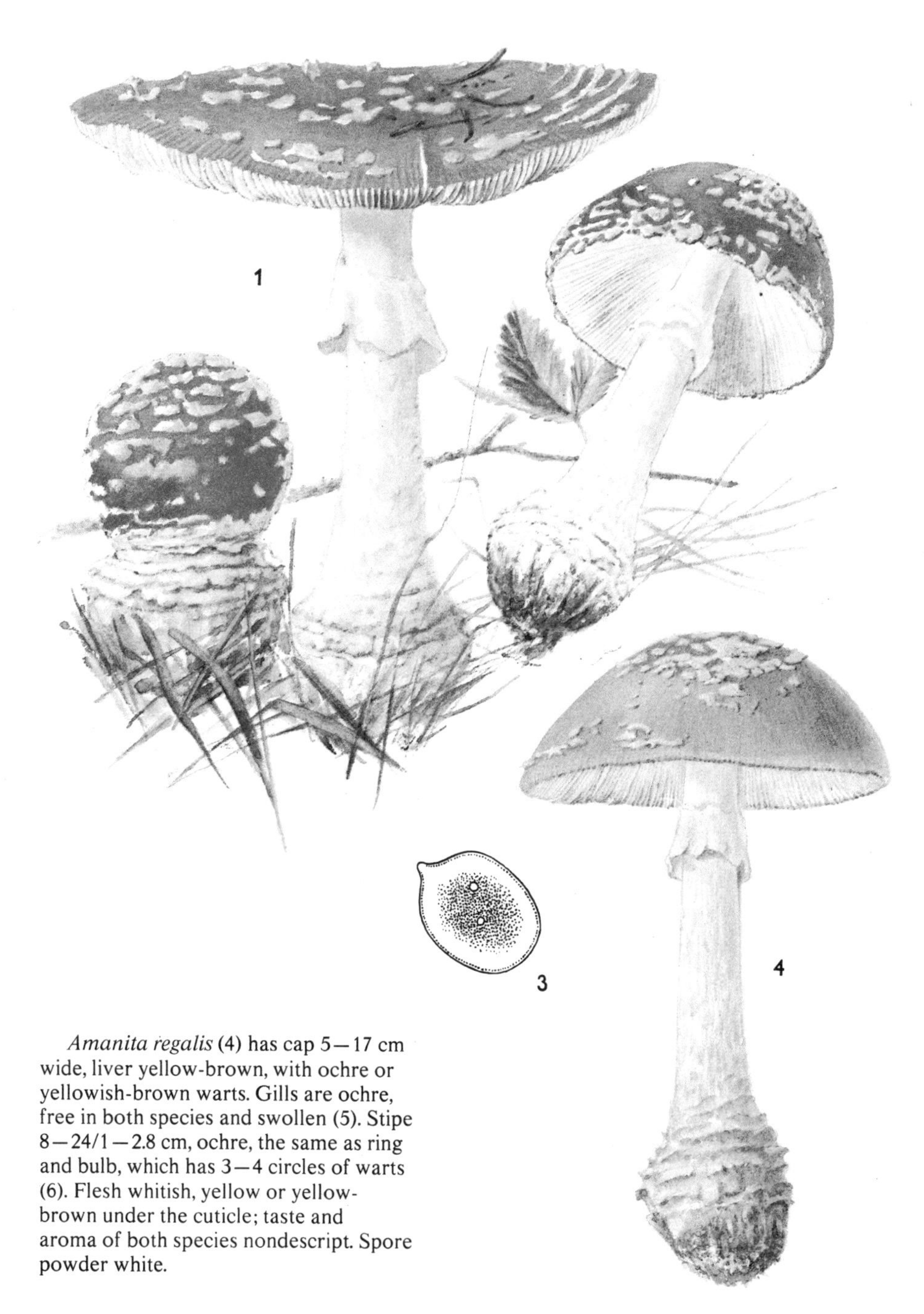


*Amanita regalis* (4) has cap 5—17 cm wide, liver yellow-brown, with ochre or yellowish-brown warts. Gills are ochre, free in both species and swollen (5). Stipe 8—24/1—2.8 cm, ochre, the same as ring and bulb, which has 3—4 circles of warts (6). Flesh whitish, yellow or yellow-brown under the cuticle; taste and aroma of both species nondescript. Spore powder white.

## Panther Cap, False Blusher

*Amanitaceae*

*Amanita pantherina* (DC ex FR.) SECR.

The Panther Cap is often the cause of very unpleasant though non-fatal poisoning. Most cases are caused by confusion with the edible *Amanita spissa.* Characteristic features of the Panther Cap include a grooved margin of the cap, a smooth ring and a rounded bulb with a collar-like volva, which often has one or two circles. In contrast, *A. spissa* has a smooth margin of the cap, a grooved ring and a beet-root-like bulb without a volva. The toxic substances in the Panther Cap are similar to those found in Fly Agaric. The effects of poisoning resemble intoxication, but they are stronger then those of Fly Agaric.

The Panther Cap is widespread throughout the temperate zone of the northern hemisphere. It grows most often in lowlands and uplands under pine, spruce, oak and beech. Its fruit-bodies appear in small groups from June to October.

*Amanita porphyria* can be easily recognized by its purple-brown cap, smooth, limp ring and a broad, subglobose bulb with traces of volva. It contains the toxin bufotenin which can cause poisoning in sensitive people. *A. porphyria* is relatively rare and occurs only in Europe, North America, northern Africa and Japan. It grows in summer and autumn only in acid soil under spruce, pine and birch.

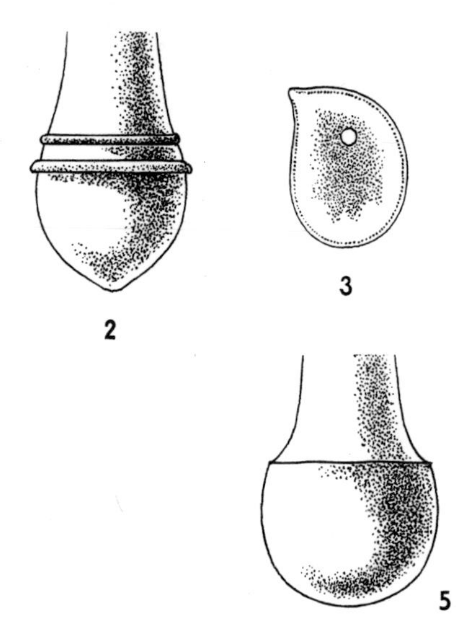


*Amanita pantherina* (1) has cap 5—12 cm wide, brown-ochre, grey-brown, with white warts and grooved margin. Gills white; stipe 6—15/0.5—1.5 cm, enlarged into a bulb, with white, delicate and smooth ring. Volva white, indistinct, thin and free, often with 1—2 circles (2). Taste and aroma nondescript. Spores (3) colourless, 10—12/7—8 μm.
*A. pantherina* can be confused with *A. spissa.*

*Amanita porphyria* (4) has cap 4 – 8 cm wide, purple-brown, sometimes with large, grey purplish warts. Gills white; stipe 7 – 10/0.7 – 2 cm, whitish or purplish. At base a wide, bordered bulb (5); grey-white ring with purplish tinge, very delicate and smooth. Volva develops as a light grey, narrow circle. Taste unpleasant, aroma of potatoes.

## Common Grisette

*Amanitaceae*

*Amanita vaginata* (BULL. ex FR.) QUÉL.

The Common Grisette and its related species of the sub-genus *Amanitopsis* differ from other *Amanita* species in lacking a ring on the stipe. Other characteristic features of this group, which includes about 10 European species, are a prominently grooved margin of the cap and in maturity a hollow stipe which becomes very fragile and breaks easily. The flesh and spore powder are white, and the taste and aroma nondescript. These species are edible, but of inferior quality. They grow from summer until autumn.

The Common Grisette is the best known species. It is probably distributed worldwide and has even been found above the arctic circle. Most often it grows in deciduous forests in acid or neutral soil under birch and beech trees. Another species, *A. umbrinolutea,* is most often found in mountainous spruce forests. The Tawny Grisette *(A. fulva)* has an entirely red-brown fruit-body, including the volva, while the related species *A. crocea* has an orange or orange-brown cap, an orange striated stipe and a completely white volva. *A. alba* is a pure white species and *A. argentea* is characterized by a silver-grey cap and stipe. The Ringed-foot *(A. inaurata)* has a brown-grey or yellow chestnut brown cap with grey warts; the scaly stipe has a similar colouring. The bulb at the base of the stipe is grey-brown inside and dirty white on the outside.

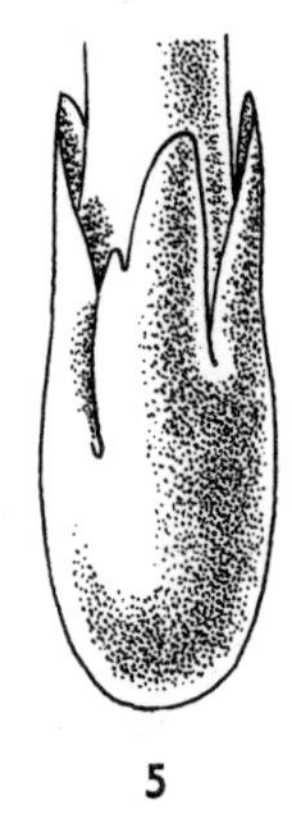


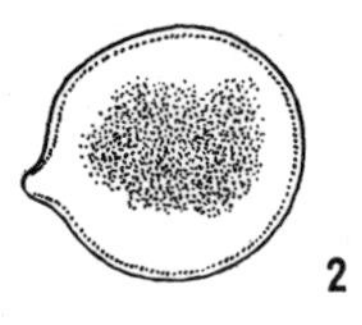


*Amanita vaginata* (1) has cap 3—10 cm wide, grey, yellow-grey. Gills white; stipe 5—15/0.8—2 cm, whitish or dirty white, hollow; volva white inside, greyish outside. Spores (2) colourless, 9—12 µm.

*Amanita umbrinolutea* (3) has cap 6—12 cm wide, ochre brown, yellow-brown with lighter margin, sometimes whitish. Gills white; stipe 8—16/0.7—1.3 cm, with transverse irregular stripes grey-brown or yellow-brown; base has white volva.

*Amanita fulva* (4) has cap 4—10 cm wide, dark red-brown. Stipe 10—15/0.7—1.5 cm, whitish or red-brown, without striations; volva entirely red-brown. Spores the same as *A. vaginata.* Volva of all species is lobed (5).

## Fawn-coloured Pluteus

*Pluteaceae*

*Pluteus atricapillus* (SECR.) SING.
Syn.: *Pluteus cervinus*

Mushrooms of the genus *Pluteus* are saprophytes, widespread throughout the world. The two most common species, *P. atricapillus* and *P. atromarginatus*, are difficult to distinguish. They have brown or black-brown caps and deep pink gills when they are mature. Young specimens have pure white gills. Only close examination reveals a black colouring of the gill edges of *P. atromarginatus*, which is caused by a dark pigment in the cystidia. The cystidia of the Fawn-coloured Pluteus are hollow. Both species grow from spring to autumn individually or in small groups on rotten stumps or fallen tree trunks, most often in damp, dark stands of virgin forests. Even during dry season when other mushrooms stop growing, *Pluteus* species fructifies because the rotten stumps retain moisture much longer than the soil. The Fawn-coloured Pluteus can be found most frequently on deciduous trees, beech and oak in particular; it is very rarely found on conifers. *P. atromarginatus*, on the other hand, grows only on conifers, particularly on spruce and fir. Both species are edible with a good taste. Both the Fawn-coloured Pluteus and *P. atromarginatus* can be confused with several related species, which are also edible and have a brown colouring.

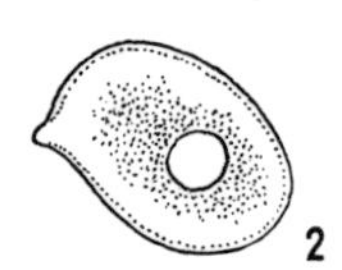


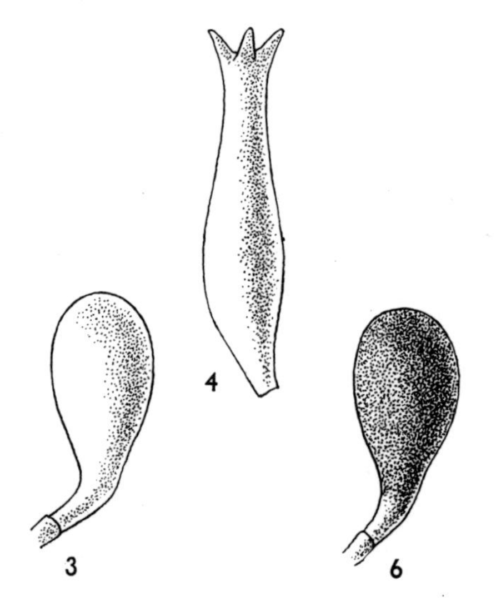


*Pluteus atricapillus* (1) has cap 4—15 cm wide, smooth, grey-brown or brown. Gills are free, whitish, later pink. Stipe 5—15/0.8—2 cm, white, longitudinally streaked with brown fibrils; flesh white. Taste and aroma mild, slightly radish-like. Spore powder pink; spores of both species (2) pale pink, in *P. atricapillus* 7—9.5/5—6 µm, in *P. atromarginatus* 6—7.5/4.5—5.5 µm. Cheilocystidia colourless (3). Pleurocystidia of both species are bottle-shaped with four hooks (4).

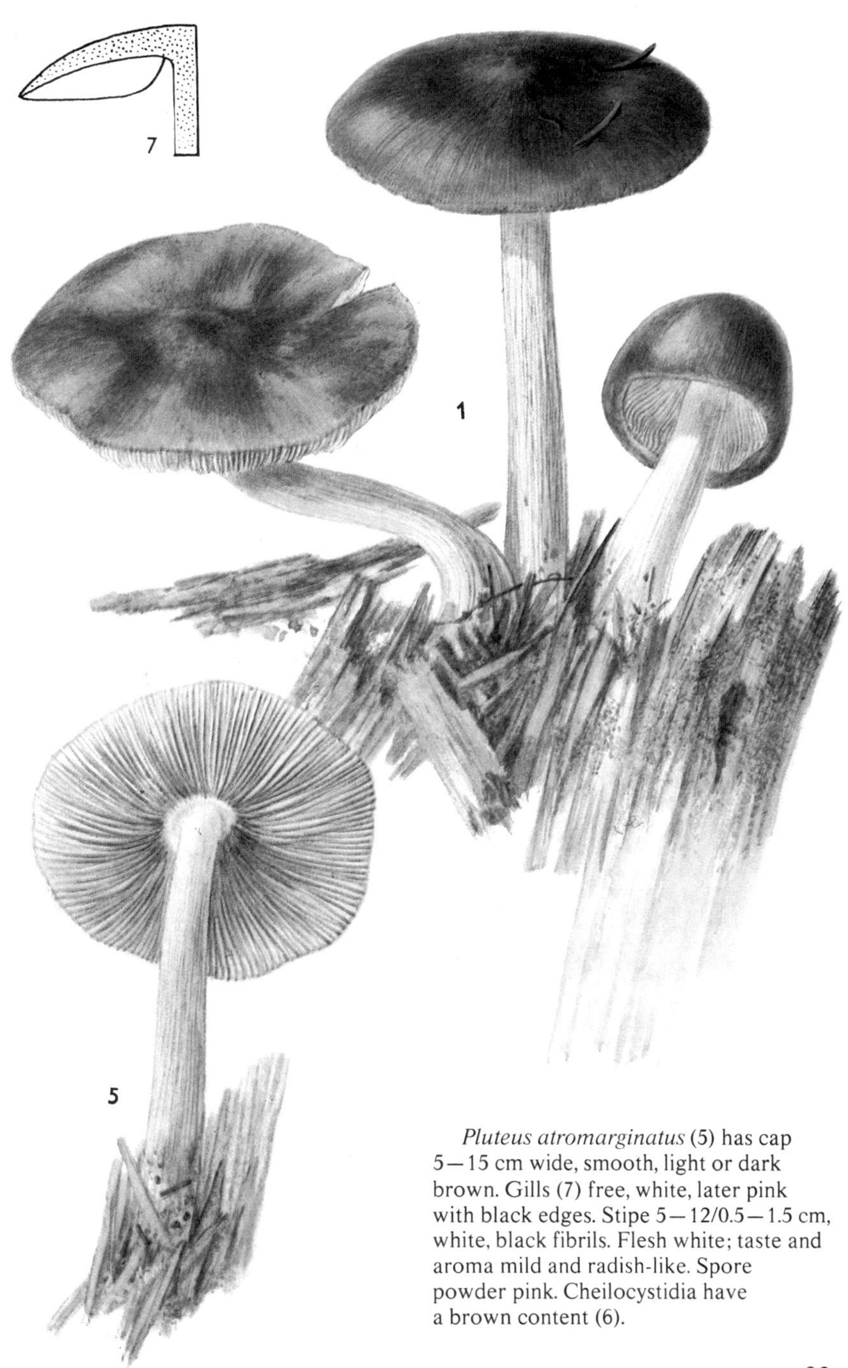


*Pluteus atromarginatus* (5) has cap 5—15 cm wide, smooth, light or dark brown. Gills (7) free, white, later pink with black edges. Stipe 5—12/0.5—1.5 cm, white, black fibrils. Flesh white; taste and aroma mild and radish-like. Spore powder pink. Cheilocystidia have a brown content (6).

## **Parasol Mushroom** *Agaricaceae*

*Macrolepiota procera* (SCOP. ex FR.) SING.
Syn.: *Lepiota procera*

The Parasol Mushroom is one of the most common species of Europe, North America, northern Africa and Asia. In some areas it is the only gill species which is trusted for use in the kitchen. The young, pestle-like fruit-bodies with enclosed caps are the most tasty. Older specimens have a tough stipe which can cause indigestion. They are rarely attacked by insects.

The Parasol Mushroom grows individually or in small groups of two or three, from June to November. It thrives in sunny, well-lit forests, in parks, gardens, pastures and meadows, preferring warm areas, lime and sandy soil. With its unique features — the cracked, coarsely scaly cap, double ring and tall, brown striated stipe — the Parasol Mushroom is very rarely confused with other species. *L. prominens* is as tall as the Parasol Mushroom but its overall colouring is lighter and the stipe has no striations; it is also edible. In North America it can be confused with the strongly poisonous *Chlorophyllum molybdites*, but the latter does not occur in Europe. *Ch. molybdites* has green gills and spore powder when mature.

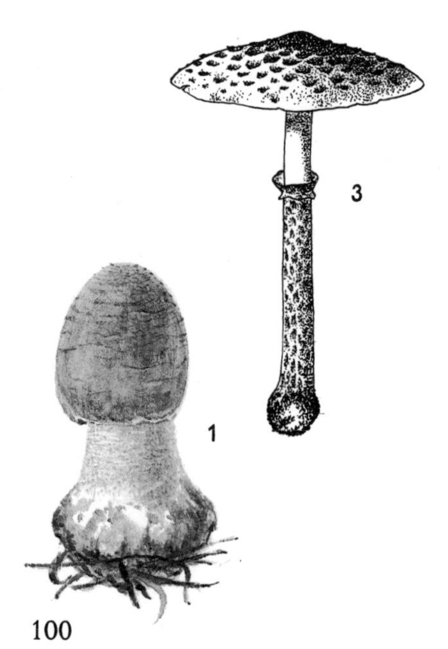


*Macrolepiota procera* (1) has cap 10—30 cm wide, beige or ochre, light brown, dark brown in the centre, smooth, later with radiating cracks or scales, woolly along the margin. Gills whitish; stipe 10—40/1.5—4.5 cm, hollow when mature, whitish, ochre, with brown stripes and coarse scales along its entire length, base enlarged into a bulb. Ring whitish, double, loose (2). Flesh white, does not change colour when exposed to the air; taste nutty, aroma pleasant. Spore powder pure white; spores colourless, 14—18/9—12 µm.

*L. mastoidea* (3) slightly resembles the Parasol but reaches only half the size. It has fine scales and an umbo on the cap; the striped brown-ochre stipe has a simple ring.

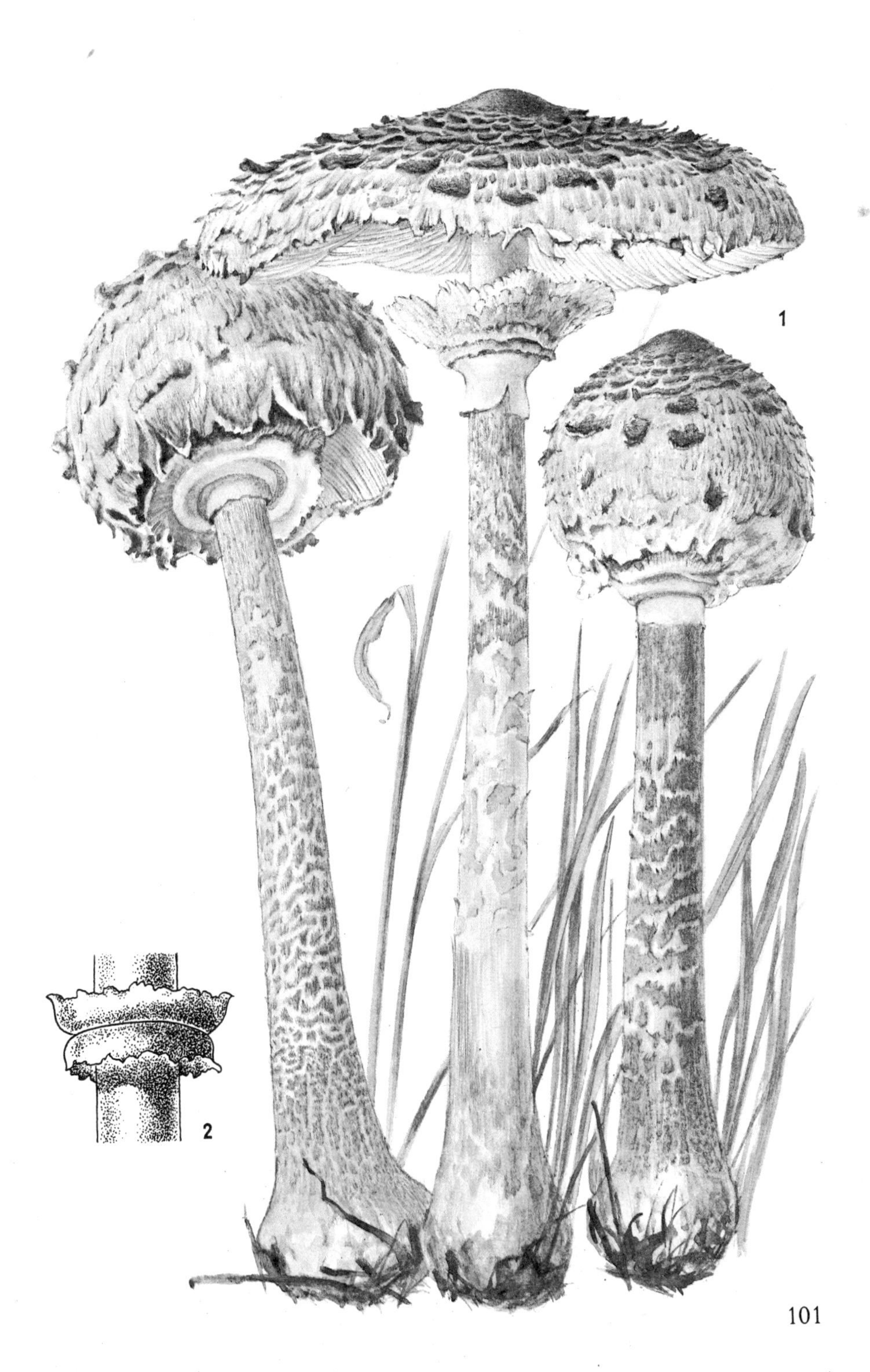
1
2

## **Shaggy Parasol** *Agaricaceae*

*Macrolepiota rhacodes* (VITT.) SING.
Syn.: *Lepiota rhacodes*

The Shaggy Parasol, a robust species found in coniferous forests, is characterized by the colour change of its flesh when cut or bruised, turning first yellow, then yellow-orange and finally red to brown. This change can put some mushroom pickers off, but the Shaggy Parasol is edible and has wide culinary use. It grows in well-lit woods on acid soil, always in groups, rows or circles of dozens of fruit-bodies, most frequently in spruce stands or under pine and less frequently in deciduous woods. The Shaggy Parasol thrives throughout the temperate zone of the northern hemisphere in the area stretching from Japan to North America. Its European variety, *hortensis,* has lighter fruit-bodies and a large, globular, bordered bulb at the base of the stipe. This variety always grows outside the forest, on compost heaps or in well-manured soil in parks and gardens. This mushroom is known to have caused several cases of slight poisoning; however, some people eat it without ill effects.

*Macrolepiota excoriata* grows in summer and autumn, usually in meadows, pastures and lawns. This rare thermophilous species can sometimes be found together with the Parasol Mushroom, and is also edible.

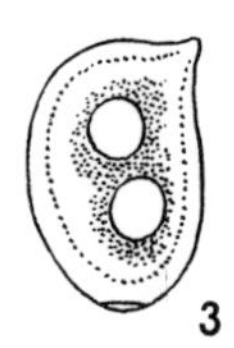


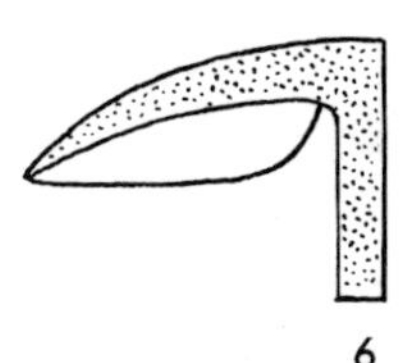


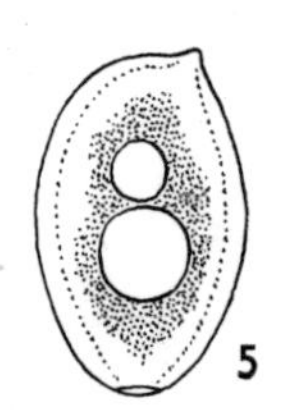


*Macrolepiota rhacodes* (1) has cap 10—15 cm wide, grey-brown, brown, with woolly filaments or overlapping scales. Stipe is 8—16/1—1.8 cm, white, red-brown when old, smooth, with a bulb; white ring. Flesh white, later turning red, particularly in the stipe (2). Taste and aroma bland. Spore powder white; spores (3) colourless, 8.5—12/6—7 μm.

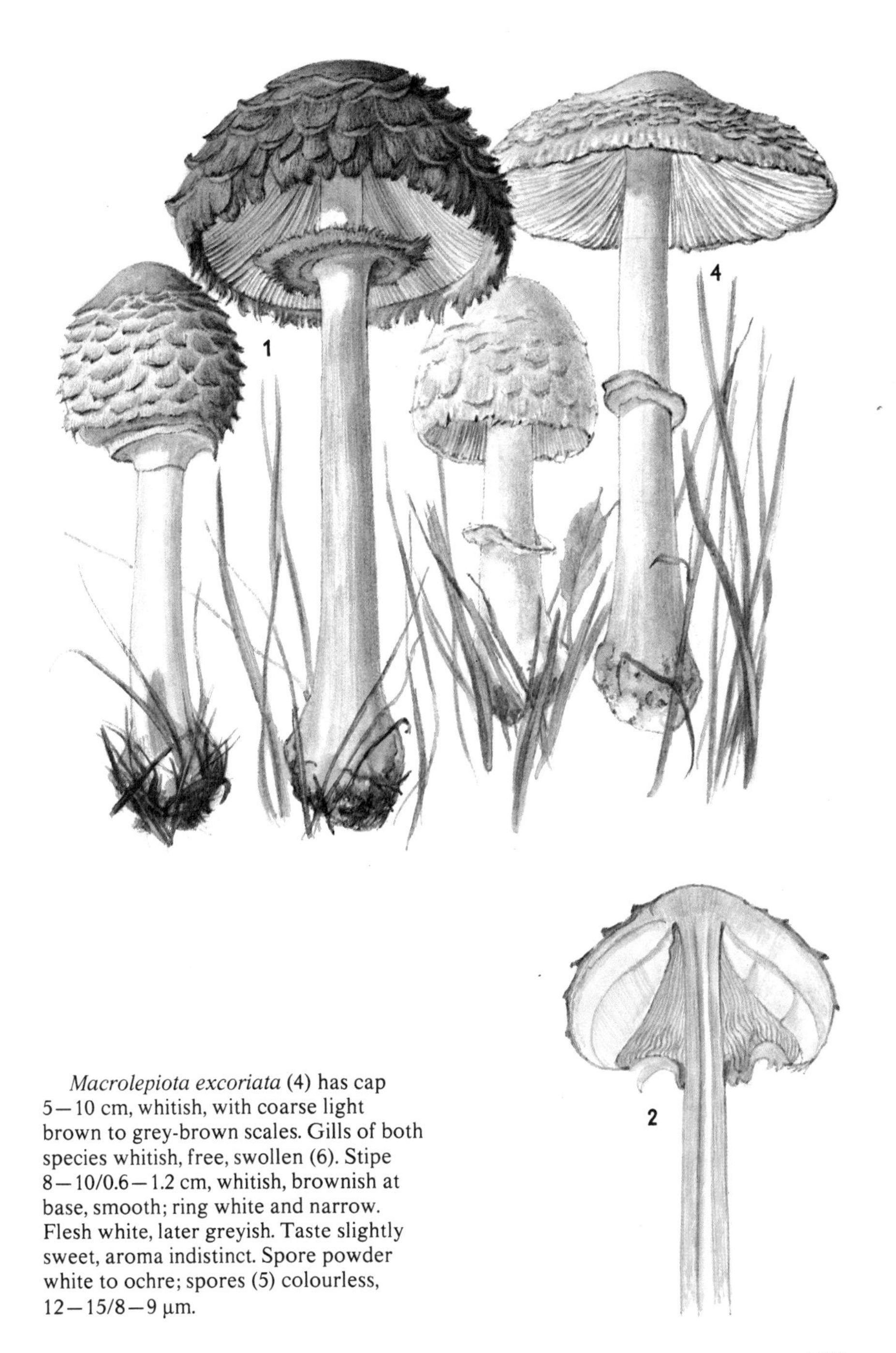


*Macrolepiota excoriata* (4) has cap 5 – 10 cm, whitish, with coarse light brown to grey-brown scales. Gills of both species whitish, free, swollen (6). Stipe 8 – 10/0.6 – 1.2 cm, whitish, brownish at base, smooth; ring white and narrow. Flesh white, later greyish. Taste slightly sweet, aroma indistinct. Spore powder white to ochre; spores (5) colourless, 12 – 15/8 – 9 µm.

## Field Mushroom

*Agaricaceae*

*Agaricus campestris* (L.) FR.
Syn.: *Psalliota campestris*

After heavy rainfall white Agaricus species thrive in pastures, gardens, meadows and fields and are collected there in large quantities. To recognize the true Field Mushroom, however, can be quite difficult. It is characterized by an almost white cap and stipe, pink to flesh-red gills; the flesh turns pink when cut. The fruit-bodies smell of freshly cut wood. The Field Mushroom shares its habitat with the poisonous *A. xanthodermus,* which is also white, but turns yellow when bruised; the flesh becoming a deep yellow in cross-section, especially at the base. The gills are not pinkish, but greyish or brown-black. The fruit-bodies smell of carbolic acid when cut or cooked.

*Agaricus bisporus* can be found in well-manured soil or directly on horse manure, in gardens and along field tracks. It has a brown scaly cap and its flesh, similar to the Field Mushroom, turns pink when cut. For its excellent flavour a pure white variety has been developed and is now widely cultivated. It is grown on agricultural refuse, on fermenting horse manure and chicken manure. Both the Field Mushroom and *A. bisporus* have a virtually worldwide distribution.

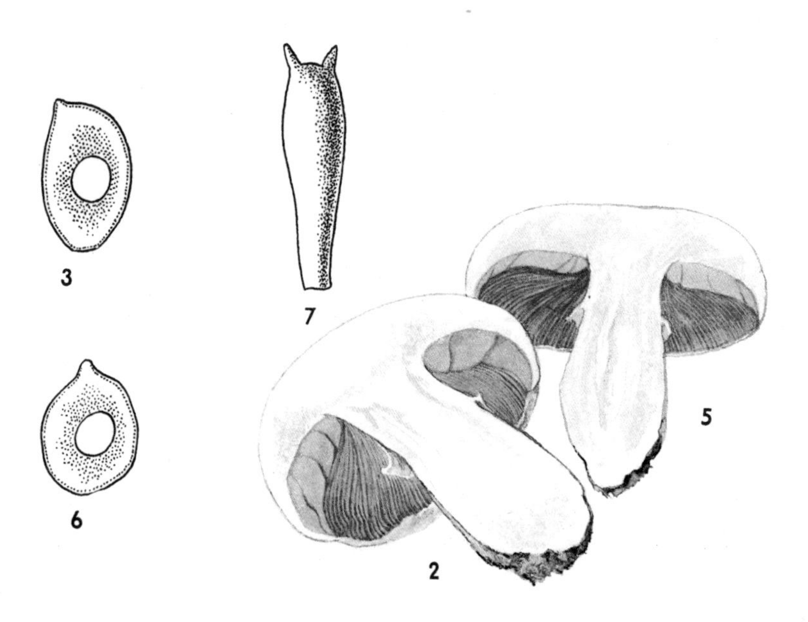

*Agaricus campestris* (1) has cap 4—8 cm wide. Stipe is 3—7/1—1.5 cm, white, slightly brownish when old, surface has ingrown fibrils. Gills pink, then flesh red and finally chocolate brown or black; white ring. Flesh whitish, turning slightly pink, especially where the cap joins the stipe (2). Taste is indistinct. Spore powder purple brown; spores (3) purple-brown, 7—8/4.5—5.5 µm.

*Agaricus bisporus* (4) has cap 5—10 cm wide, greyish brown, with brown fibrils and scales. Gills white, soon becoming pink, later red-brown, with whitish blades. Stipe 3—5/1—1.5 cm, whitish, often with brown patches; white ring. Flesh whitish, later pink (5). Taste indistinct, aroma of fresh cut wood. Spore powder brown; spores (6) brown, 4—7.5/4—5.5 µm. Basidia have two short stalks (7).

## Yellow Stainer

*Agaricaceae*

*Agaricus xanthodermus* GEN.
Syn.: *Psalliota xanthoderma*

The Yellow Stainer is a toxic species which is responsible for poisonings every year because of its close similarity to most edible Field Mushrooms. The effects are not fatal, but include vomiting and intestinal disorder. Anybody who collects Field Mushrooms for use in the kitchen should be able to distinguish them from the Yellow Stainer. Its main features include the slight yellowing of the entire fruit-body when it is bruised, and the silky glossy stipe, which turns an intensive yellow at the base. It has an unpleasant aroma of carbolic acid, which becomes stronger during cooking. The Yellow Stainer can be found from June to October in meadows, pastures, gardens, parks and also in coniferous and deciduous forests. It is particularly abundant after heavy rainfall. Its distribution is almost worldwide.

Like *A. xanthodermus*, *A. meleagris*, *A. placomyces* and *A. phaeolepidotus* are poisonous. They all smell of carbolic acid and turn deep yellow at the base of the stipe. They grow in summer and autumn outside the forest. The differences between these species are so minute that they are often considered to be one species.

*A. ammophilus*, which inhabits sandy soil of the European coastline, has a white cap with a grey-pink tint, without scales. The fruit-bodies reach about half the size of *A. xanthodermus*.

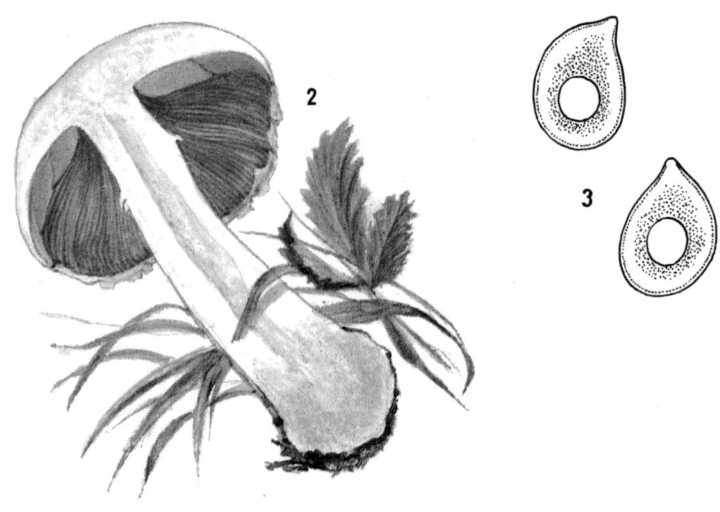

*Agaricus xanthodermus* (1) has cap 5—13 cm wide, smooth, white, yellow or greyish, turning yellow when bruised. Gills remain pale pink or greyish for a long time, later brown-red. Stipe 6—12/1.2—1.5 cm, white, with a silky sheen, yellow or greyish with a white ring. Flesh white, turning faint yellow; base of the stipe deep chrome yellow (2). This and the following species smell of carbolic acid. Spore powder brown-black; spores of both species (3) dark brown, 5—7/3—4 μm.

*Agaricus meleagris* (4) has cap 5—12 cm wide, with grey-brown to brown-black small scales, darker in the centre, turning yellow when bruised. Stipe 6—10/1—1.2 cm, whitish or brownish, yellowing when bruised, with a white ring. Flesh whitish; turns chrome yellow at base (5).

## **Wood Mushroom** *Agaricaceae*

*Agaricus sylvaticus* SCHFF. ex SECR.
Syn.: *Psalliota sylvatica*

Although most Agaricus species live outside the forest, *Agaricus sylvaticus* and the Prince *(A. augustus)* are typical woodland mushrooms. They both grow only in coniferous, mostly spruce, forests in places rich in nitrogen. *A. sylvaticus* is a common mushroom which thrives throughout the temperate zone of the northern hemisphere, growing in groups from June to October. Its cap is never white, but covered with dark brown small scales. If fresh, the entire fruit-body turns carmine red when bruised. *A. sylvaticus* tastes excellent; the young fruit-bodies with grey-red gills being the most delicate. Several varieties, all edible and of excellent flavour, can be recognized by the colour of the cap and of the scales. A related species, *A. langei*, grows in coniferous forests; it has a similar colouring, but is larger, and the flesh turns vivid red when bruised. It is also edible.

The Prince is one of the rarest, largest and most handsome *Agaricus* species, growing in small, prominent groups of stout, fleshy fruit-bodies. Its flavour is also most delicate. It is chracterized by its large size, yellowish cap with dark brown scales and slight yellowing when bruised. A closely related species, *A. perrarus*, also inhabits coniferous forests; its cap has loose ochre or yellow-brown scales and it is also edible.

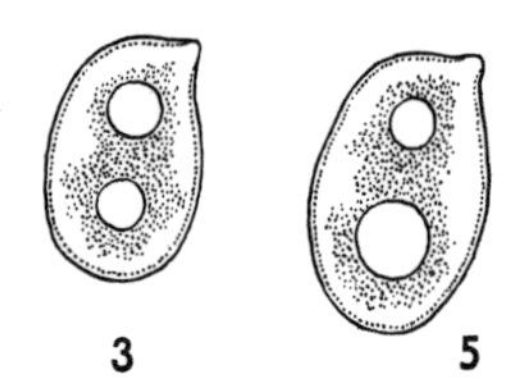


*Agaricus sylvaticus* (1) has cap 5—10 cm wide. Gills light grey-red to brown-grey, later chocolate brown. Stipe 6—12/1—1.5 cm, whitish to grey, smooth; ring white. Flesh whitish, later light carmine to purple (2). Taste indistinct, aroma faintly spicy. Spore powder purple-brown, spores (3) purple-brown, 4.5—6/3—3.5 µm.

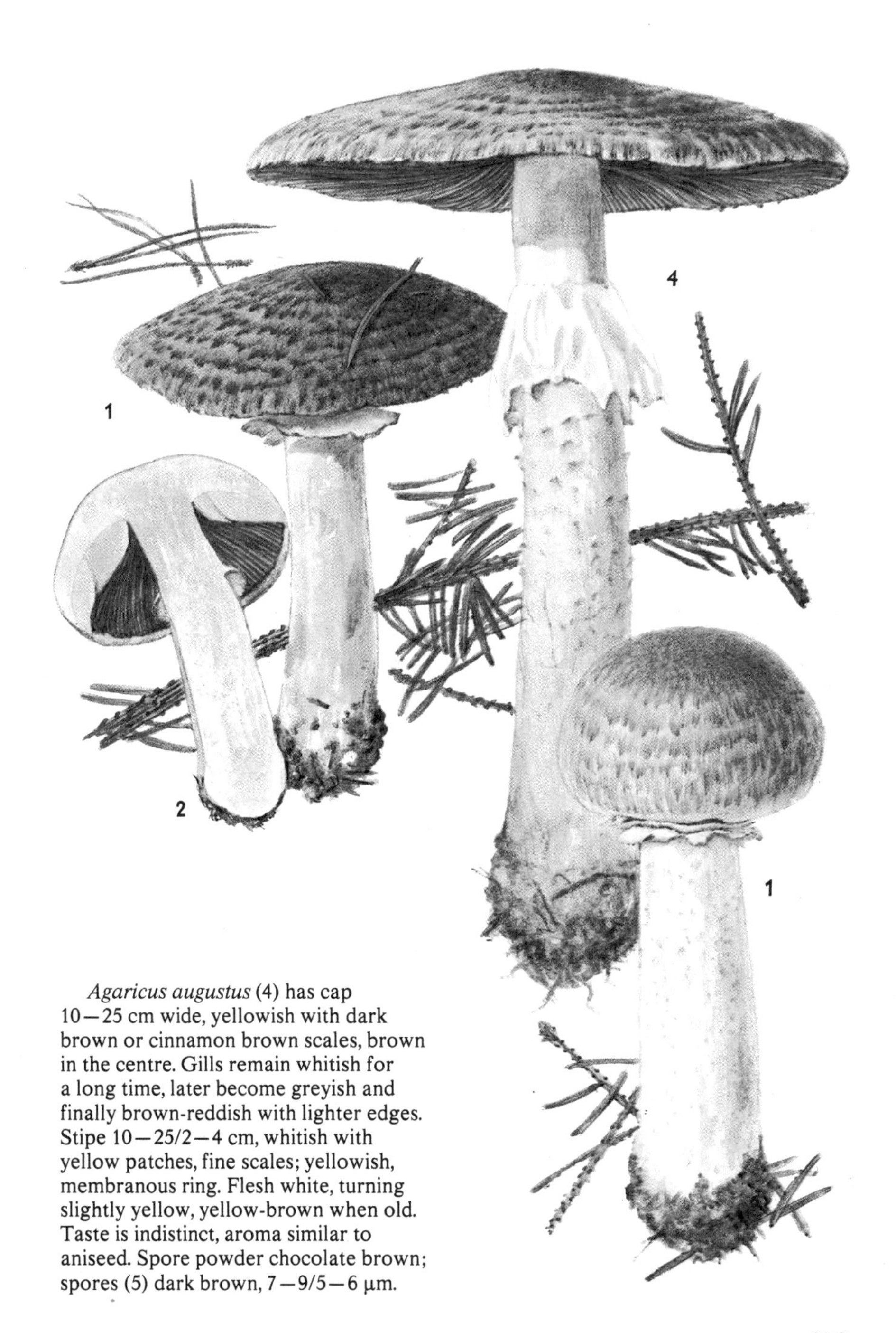


*Agaricus augustus* (4) has cap 10—25 cm wide, yellowish with dark brown or cinnamon brown scales, brown in the centre. Gills remain whitish for a long time, later become greyish and finally brown-reddish with lighter edges. Stipe 10—25/2—4 cm, whitish with yellow patches, fine scales; yellowish, membranous ring. Flesh white, turning slightly yellow, yellow-brown when old. Taste is indistinct, aroma similar to aniseed. Spore powder chocolate brown; spores (5) dark brown, 7—9/5—6 μm.

## **Horse Mushroom** *Agaricaceae*

*Agaricus arvensis* SCHFF. ex FR.

The Horse Mushroom appears soon after rain on grazing land, meadows and in gardens, but rarely in forests. It grows in small groups from June to October in the entire temperate zone of both hemispheres. It can be recognized by the yellowing of the bruised cap and stipe, by the flesh grey gills of the adult mushroom and by its double ring.

The Horse Mushroom is closely related to several species, including *A. sylvicola,* which grows from August to October in forests only. It differs from *A. arvensis* by its single ring and smaller spores. *A. abruptibulbus* has a stipe with a bulbous base and a single ring. It appears mainly in coniferous forests from June to November. *A. macrosporus* has large fruit-bodies and its flesh turns a faint red when cut. It can be found in meadows and along the edges of woods. All the above species are edible and have a pleasant flavour. They are commonly confused with *A. xanthodermus,* which grows in the same season and shares the same habitat. Though all the former species turn yellow when bruised, the silky, glossy stipe of *A. xanthodermus* becomes an especially deep chrome yellow at the base; this can be seen in a cross-section of the mushroom. *A. xanthodermus* smells unpleasantly of carbolic acid, while the Horse Mushroom and its related species have a pleasant aniseed aroma. Less commonly, but sometimes with fatal results, the Horse Mushroom and its relatives are confused with the poisonous White Spring Amanita or the Destroying Angel.

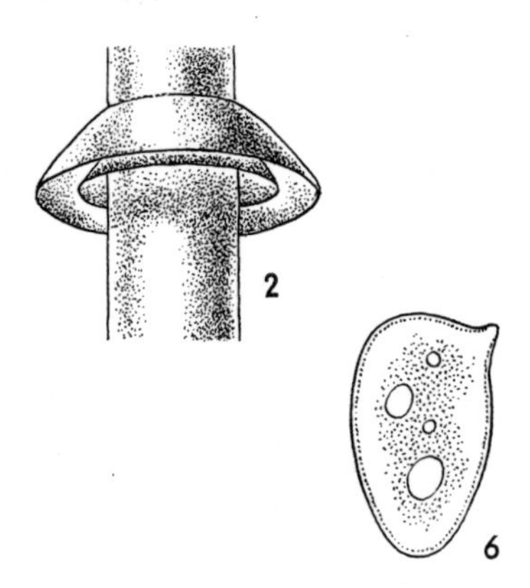

*Agaricus macrosporus* (4) has cap 10—30 cm wide; stipe 5—10/3—4 cm, turns yellow when bruised. Flesh in cross-section becomes lightly red (5). Spores (6) 10—12/6.5—7 μm.

*Agaricus arvensis* (1) has cap 7—15 cm wide, white, turning yellow when bruised, old fruit-bodies are yellowish or light ochre; smooth or delicately scaly. Gills remain yellowish for a long time, later becoming a pale flesh colour and finally chocolate brown when old. Stipe 8—13/1.5—3 cm, whitish, turning yellow when bruised, with double white to yellow ring (2). Flesh whitish, ochre when old; spores purple-brown, 7—8/3—4 μm. Careless mushroom pickers sometimes confuse the Horse Mushroom with the deadly poisonous Death Cap.

*Agaricus abruptibulbus* (3) resembles *A. arvensis* in all features except for its loose, flat bulb at the base of the stipe, single ring and small spores.

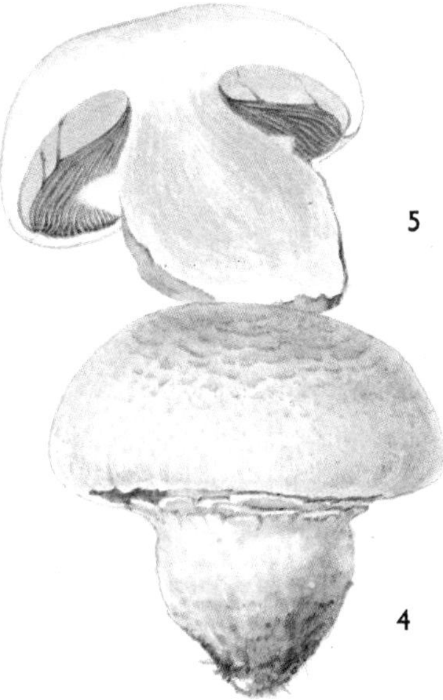

# Shaggy Cap, Lawyer's Wig

*Coprinaceae*

*Coprinus comatus* (MÜLL. ex FR.) S. F. GRAY

*Coprinus* species appear soon after rain, often on the same day, and they develop very quickly, at times reaching full size within an hour. However, the life of the fruit-bodies is very short; they begin to die off after several hours and as a result of the activity of their own enzymes they undergo a process of self-digestion, during which they change into a thick black substance. They live as saprophytes in soil, manure or wood and have a nearly worldwide distribution.

The Shaggy Cap can be found in well-manured, grassy places, in gardens and along the woodland edges from spring to autumn. It grows individually or in sparse clusters and is edible when young. The related *C. ovatus* has an ovoid rather than cylindrical cap, and its gills change colour from white to black, without turning pink, as with the Shaggy Cap.

The Common Ink Cap *(C. atramentarius)* grows in summer and autumn in thin clusters at the base of stumps and trunks of deciduous trees and also in well-manured gardens and along field tracks. It is poisonous if alcohol is consumed during the meal or even several days after it. This mushroom contains a substance, coprine, which interferes with alcohol metabolism at the acetaldehyde stage and raises its concentration to a dangerous level. Therefore the poisoning is caused not directly by poisonous substances in the mushroom, but by the acetaldehyde. In summer and autumn the Glistering Coprinus *(C. micaceus)* forms thick clusters on dead trunks, roots and stumps of deciduous trees. The young mushrooms are edible, but it is likewise not advisable to drink alcohol during or after the meal.

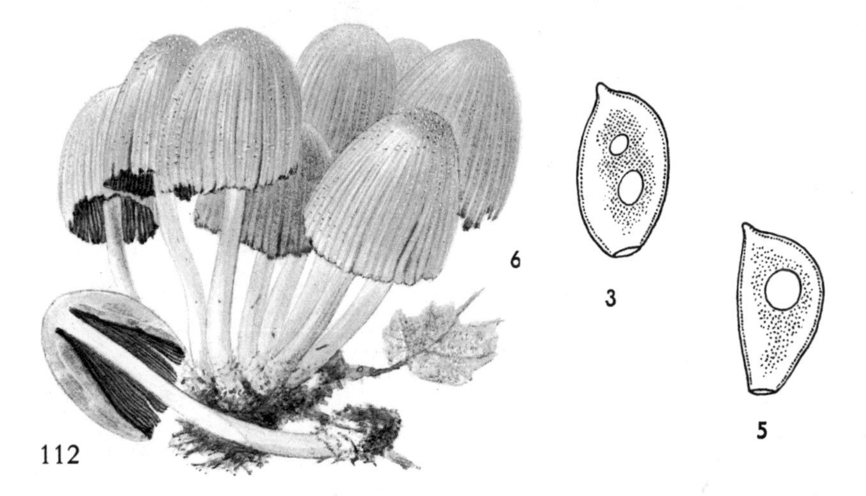

*Coprinus comatus* (1) has cap 3—6 cm wide, cylindrical, white, coarsely scaly, with ochre top. Gills white, later pink, dissolving into black mass. Stipe 10—20/0.7—1.5 cm, white, with a ring. Flesh of the cap is white, later red to black; flesh of the stipe is white (2). Spore powder black; spores (3) black, 12—16/7—8 µm.

*Coprinus atramentarius* (4) has cap 3—8 cm wide, grey, grey-brown, grooved. Gills white, later black; stipe 8—12/1—2 cm. Spore powder black; spores (5) black, 7.5—10/5—5.5 µm.

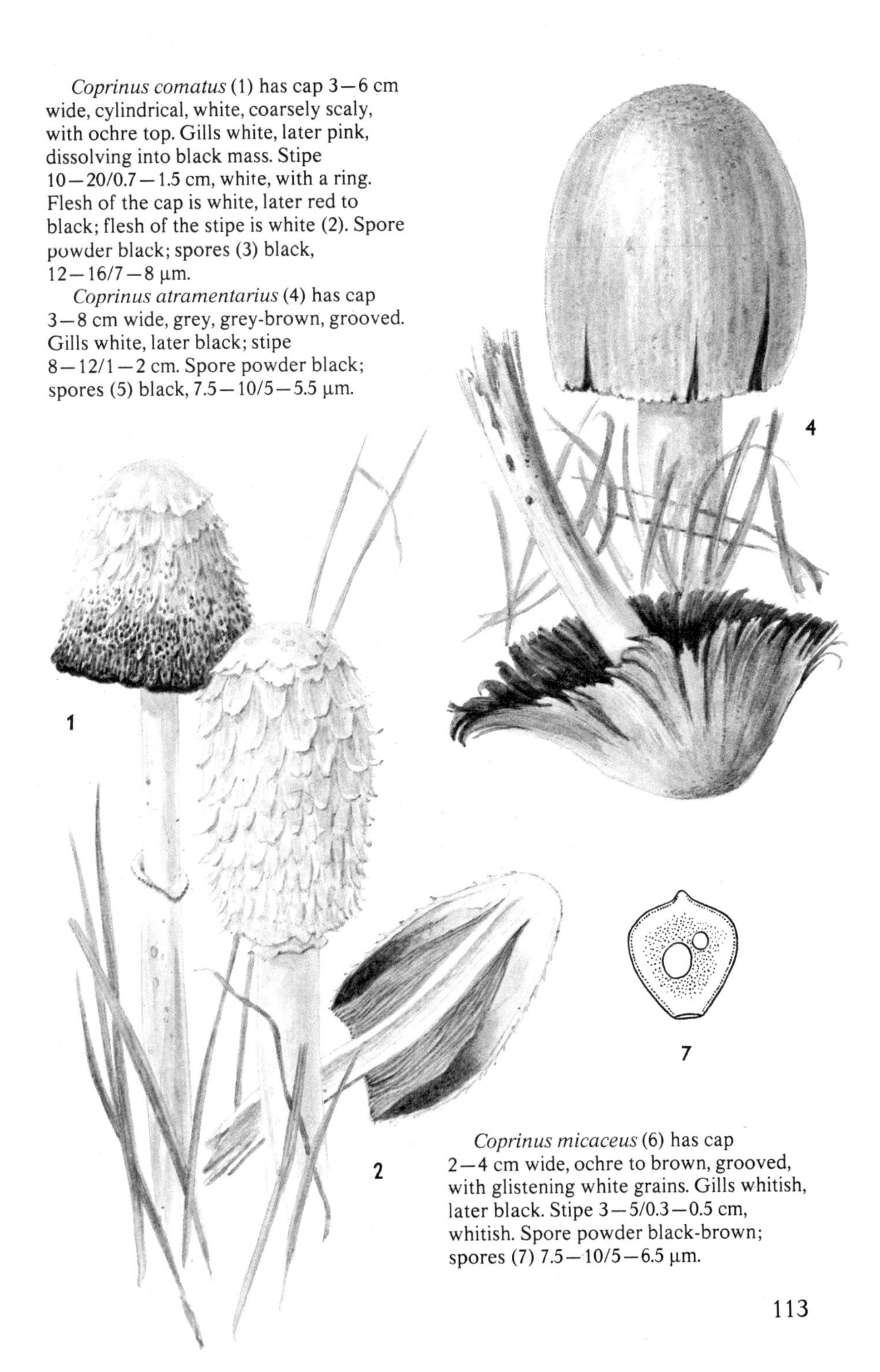


*Coprinus micaceus* (6) has cap 2—4 cm wide, ochre to brown, grooved, with glistening white grains. Gills whitish, later black. Stipe 3—5/0.3—0.5 cm, whitish. Spore powder black-brown; spores (7) 7.5—10/5—6.5 µm.

## Verdigris Agaric

*Strophariaceae*

*Stropharia aeruginosa* (CURT. ex FR.) QUÉL.

Mushrooms of the genus *Stropharia* are small saprophytes, which have similar habitats to *Agaricus* species. Some *Stropharia* species prefer well-manured places outside forests, while others grow in forests on rotten wood. Some kinds grow directly on horse or cattle manure. All 18 European species have moist, slimy caps and brown-purple or black-purple spore powder. *S. rugosoannulata* is cultivated in bulk in some countries, in the same way as Field Mushrooms.

The Verdigris Agaric grows in groups or clusters on dead trunks or stumps of coniferous trees, on spruce, pine or fir trees, and less commonly on dead deciduous trees. It is abundant in summer and autumn in lowlands as well as highlands. In Europe the Verdigris Agaric is considered edible, but of inferior quality. Before cooking the slimy cuticle of the cap should be peeled off. In the United States of America several cases of poisoning have been reported and therefore this mushroom is classified as poisonous.

A similar though less common species, *S. cyanea,* grows outside forests in summer and autumn on grass, in meadows, pastures and lawns in Europe and North America. It is edible, but its flavour is inferior.

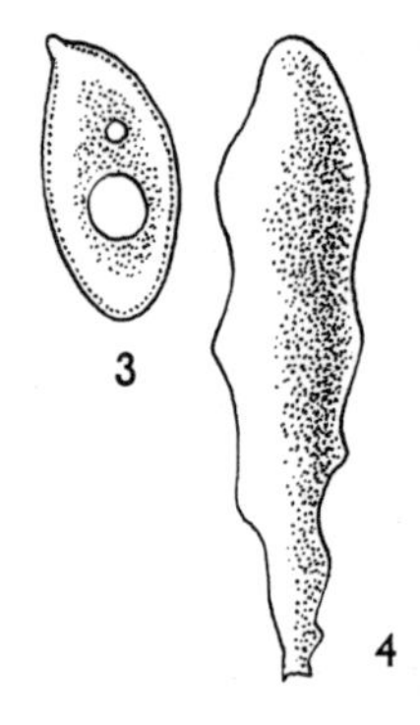


*Stropharia aeruginosa* (1) has cap 3—10 cm wide, verdigris green with ochre patches. Gills whitish, later purple-grey (2). Stipe 4—12/0.8—2 cm, slimy, pale blue or pale green, with whitish scales or flakes under the whitish, often disappearing ring. Flesh greenish to bluish; taste is lightly radish-like, aroma indistinct. Spores (3) dark brown, 7.5—9/4.5—5 µm. Cystidia on gill edges are wavy (4); in *S. cyanea* they are bottle-shaped (8).

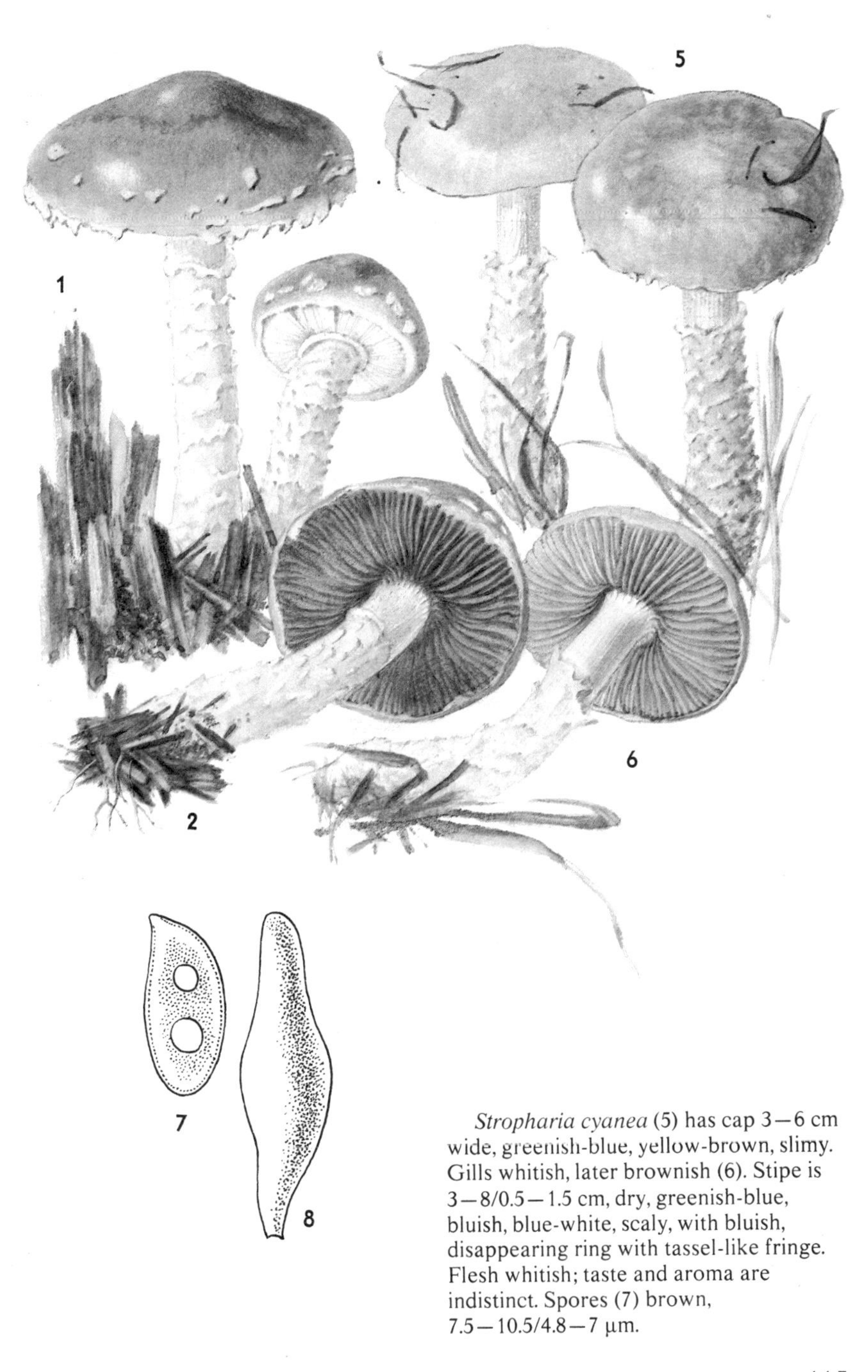


*Stropharia cyanea* (5) has cap 3–6 cm wide, greenish-blue, yellow-brown, slimy. Gills whitish, later brownish (6). Stipe is 3–8/0.5–1.5 cm, dry, greenish-blue, bluish, blue-white, scaly, with bluish, disappearing ring with tassel-like fringe. Flesh whitish; taste and aroma are indistinct. Spores (7) brown, 7.5–10.5/4.8–7 µm.

## **Sulphur Tuft, Clustered Woodlover** *Strophariaceae*

*Hypholoma fasciculare* (HUDS. ex FR.) KUMM.
Syn.: *Naematoloma fasciculare*

The Sulphur Tuft grows from spring to late autumn in large tufts of up to fifty fruit-bodies with their stipes fused together, on stumps or at the base of trunks, and also on roots in the ground. The Sulphur Tuft grows on various conifers and deciduous trees — on pine, spruce, beech, birch and oak and in Africa also on cedar. It is very common in Europe, North America, northern Africa and Asia and can be also found in Australia and southern Africa. It is classified as a poisonous species, as it causes stomach and intestinal disorders. The Sulphur Tuft has an unpleasant, bitter taste but is sometimes confused with certain autumn *Hypholoma* species, such as *H. capnoides* and *H. sublateritium,* or with the Honey Fungus and the Changing Pholiota. The Sulphur Tuft is distinguished from the above-mentioned species by the sulphur yellow flesh of its cap, yellow-green gills and bitter taste.

*Pholiota lenta,* a related but rarer species, grows on stumps and roots of coniferous and deciduous trees or on fallen foliage. It appears in summer and is most abundant in autumn. This species does not form clusters, but its individual fruit-bodies congregate in small groups. *Ph. lenta* is an edible mushroom, but it is less common.

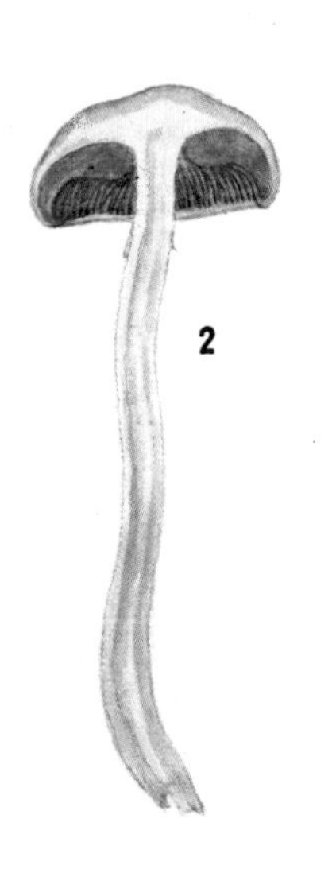


*Hypholoma fasciculare* (1) has cap 3—6 cm wide, sulphur yellow, red-brown or rusty red in the centre. Gills sulphur yellow, later yellow-green to purple-black. Stipe is 3—7/0.5—1 cm, hollow, yellow, brown at the base. Flesh sulphur yellow (2); taste is bitter, aroma indistinct. Spore powder brown-purple; spores (3) brown, 6—8/3.5—4.5 µm. Cystidia (4) have yellow content.

*Pholiota lenta* (5) has cap 4—7 cm wide, slimy, whitish, dingy yellow, yellow-brown, often with scaly remnants of veil. Gills whitish, later dingy yellow. Stipe is 5—8/0.6—1.2 cm, whitish, brownish at base, covered with white scales. Flesh pale yellowish; taste resembles radishes, aroma distinct. Spore powder ochre; spores (6) pale yellow-brown, 6—7/3—4 µm. Cystidia spindle-shaped, elongated at the apex (7).

1
5
3
4
6
7

## *Hypholoma capnoides* FR. ex FR. — *Strophariaceae*
## Syn.: *Naematoloma capnoides*

Except for several species which grow on peat and moss, most mushrooms of the genus *Hypholoma* are wood-inhabiting saprophytes. *H. capnoides,* a typical wood saprophyte, forms clusters on stumps or roots concealed in the ground. It grows only in coniferous forests, both in lowlands and high in mountains, being particularly abundant in spruce forests. It is widespread throughout the temperate zone of the northern hemisphere and it can be picked from spring to autumn and even during mild winters. This excellent edible mushroom is characterized by its poppy seed-like blue-grey gills, yellow-brown cap, whitish flesh and pleasant flavour.

*H. capnoiaes* can be confused with several similar, but bitter-tasting species. These include the poisonous Sulphur Tuft with its yellow-green gills, sulphur-yellow bordered cap and sulphur yellow flesh and *H. sublateritium,* which has yellow-brown gills and a brick-red cap. The latter forms clusters on stumps of beech and oak trees in deciduous woods throughout summer and autumn. A similar species with a slightly bitter taste is *H. radicosum,* which has an earth yellow cap with a brown centre and silky, flaky surface; the margin of the cap has tassel-like veil remnants and the gills are earth brown with white edges. The long rooted stipe has the same colouring as the stipe and is covered with scales and flakes. This rare species grows individually on the roots of coniferous trees in submountainous and mountainous regions.

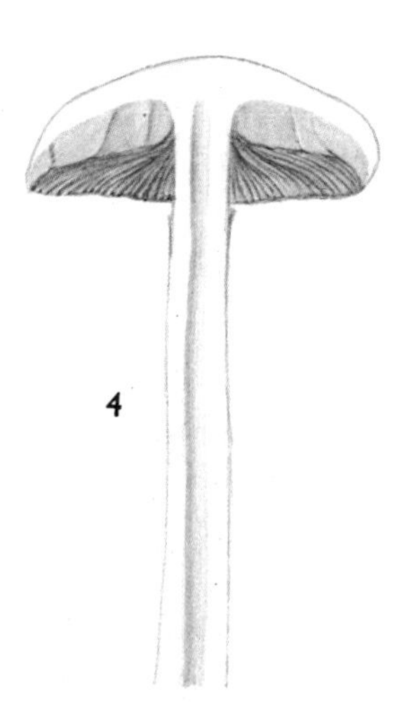


*Hypholoma capnoides* (1) has cap 2—6 cm wide, yellow-brown, orange-brown. Gills whitish when young, later smoky, poppy seed-like blue-grey. Stipe 3—10/0.4—0.8 cm, whitish at the top, brown-rusty at the base, sometimes with remnants of the ring. Flesh whitish to light yellow (2); taste and aroma pleasantly fungoid. Spore powder grey-blue; spores brown, 7—9/4—5 μm.

*Hypholoma sublateritium* (3) has cap 3–10 cm wide, brick red, rusty. Gills yellowish, later yellow-brown to olive black. Stipe 7–12/0.6–1 cm, ochre, rusty brown at base with cobweb-like remnants of veil. Flesh whitish (4); taste bitter, aroma indistinct. Spore powder purple-brown; spores the same colour, 6–8/3–4 μm.

## Shaggy Pholiota, Scaly Cluster Fungus *Strophariaceae*

*Pholiota squarrosa* (BATSCH ex FR.) KUMM.

Mushrooms of the genus *Pholiota* are wood-inhabiting saprophytes or parasites. They all have a veil when young, which later breaks; its remnants hang along the margin of the cap or remain on the stipe in the form of a ring.

The most common species of the genus, the Shaggy Pholiota *(Ph. squarrosa),* is widespread in Europe, North America and Japan. It grows in summer and autumn in clusters on roots, stumps or along the base of trunks of beech, apple or spruce trees. It is edible, but of inferior quality, as its fruit-body is tough and bittery in taste. In autumn this mushroom can be confused with the Honey Fungus, though the latter is not yellow and has no coarse scales. The Shaggy Pholiota resembles several related species, including the inedible *Ph. flammans,* which has non-slimy fruit-bodies entirely covered with lemon yellow scales and grows on coniferous stumps. The Fat Pholiota *(Ph. adiposa),* which is edible, has slimy fruit-bodies with sparse appressed scales, unlike those of *Ph. squarrosa,* which are remote. It grows on deciduous trees.

The Destructive Pholiota *(Ph. destruens),* a stout and fleshy autumn species, lives on poplars, either on the stumps or felled trunks, on worked wood or even on living trees. It can also be found on willow trees. It appears individually or in small clusters. It is inedible because of its bitter taste.

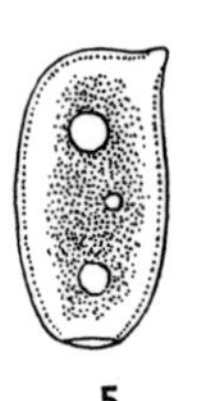


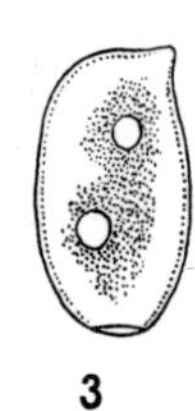


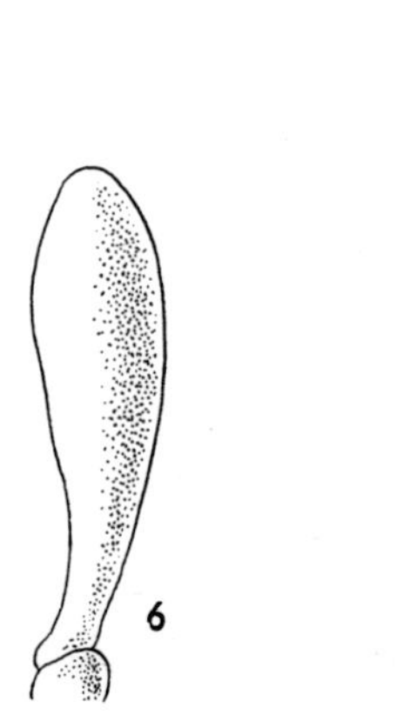


*Pholiota squarrosa* (1) has cap 4—10 cm wide, stipe 5—12/1—1.5 cm. Fruit-bodies are straw yellow, covered with dry, rusty brown, upward-turned scales. Gills attached in curves (2), yellow, later rusty. Flesh yellowish; taste slightly bitter, radish-like, aroma also resembles radishes. Spore powder dark brown; spores (3) are ochre-brown, 6—8/3.5—4 μm.

*Pholiota destruens* (4) has cap 5—20 cm wide, ochre, beige to brownish with wooly, light brown scales, slimy. Gills ochre, later brown. Stipe 3—10/2—3 cm, whitish to ochre, with light scales, enlarged at the base and rooting. Flesh whitish; taste bitter, very strong aroma. Spore powder dark rusty brown; spores (5) olive yellow, 7.5—9/5—6 µm. Cystidia of both species have the same appearance (6).

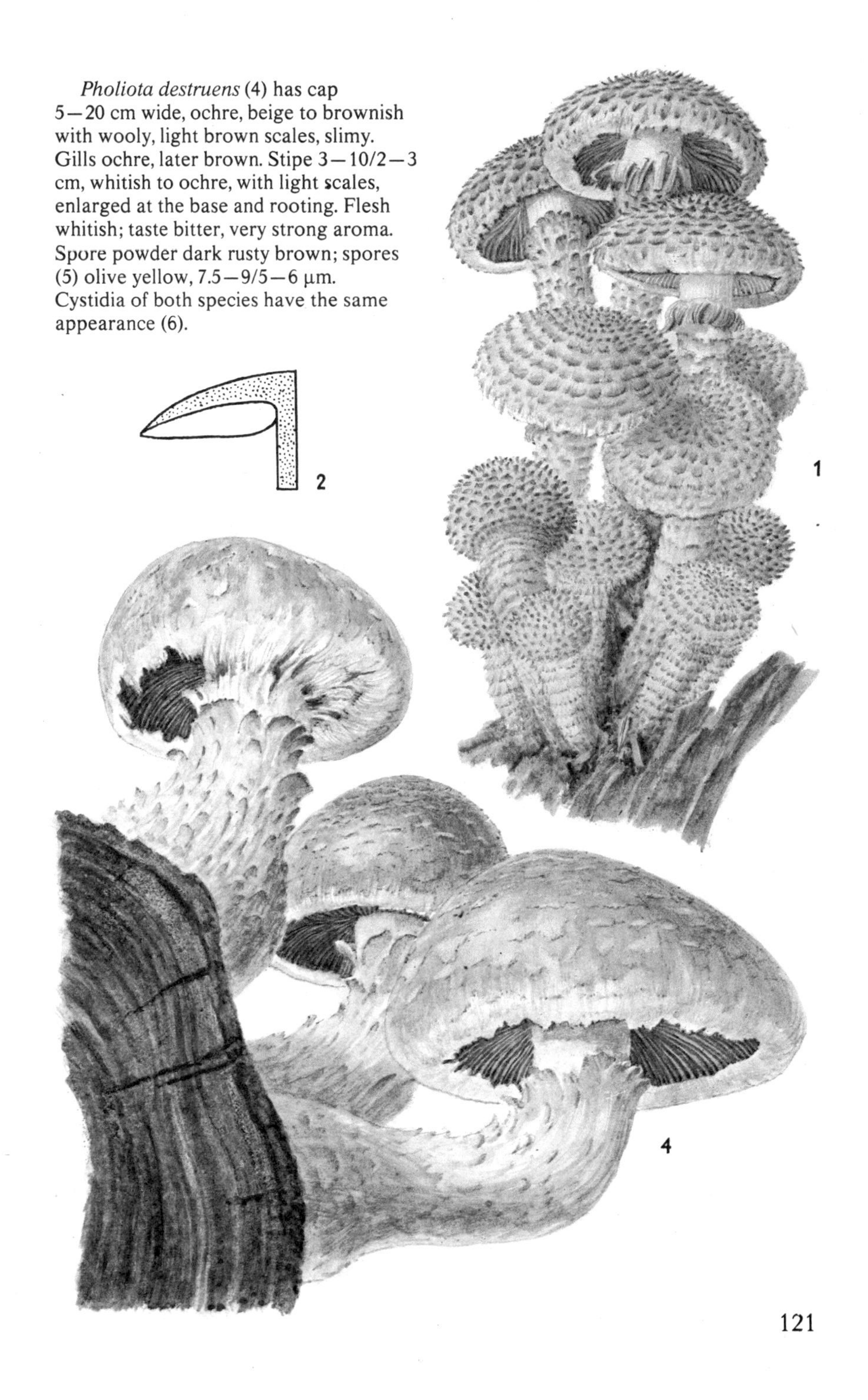

## Red-staining Inocybe
*Inocybe patouillardii* BRES.

*Cortinariaceae*

*Inocybe patouillardii* was described in 1903 by an Italian priest, G. Bresadola, and named by him in honour of the famous French mycologist, N. Patouillard. This relatively rare species can be found only in warm areas on lime-rich soil. It grows in mycorrhizal association with lime, beech, oak and hornbeam in old parks, gardens and also in deciduous forests, appearing in May and lasting until July. *Inocybe patouillardii* and the related *I. godeyi* are both characterized by their fruit-bodies which turn red when touched or when old. *I. godeyi* grows in deciduous forests on limey soil. It is quite common and is often confused with *I. patouillardii.* However, *I. godeyi* does not appear until summer; is smaller and its stipe is enlarged into a bulb at the base.

All *Inocybe* species, which are represented in Europe by approximately 140 species, are considered poisonous. They contain the toxic alkaloid muscarine; the same poison is found in some *Amanita* species, but is present in *Inocybe* species in much greater quantities. The course of poisoning is very rapid: the first symptoms of vomiting, diarrhoea, muscle weakness and partial loss of sight appear after 15 to 30 minutes. If atropine is administered quickly, the patient is cured in two days. Untreated cases of poisoning last much longer.

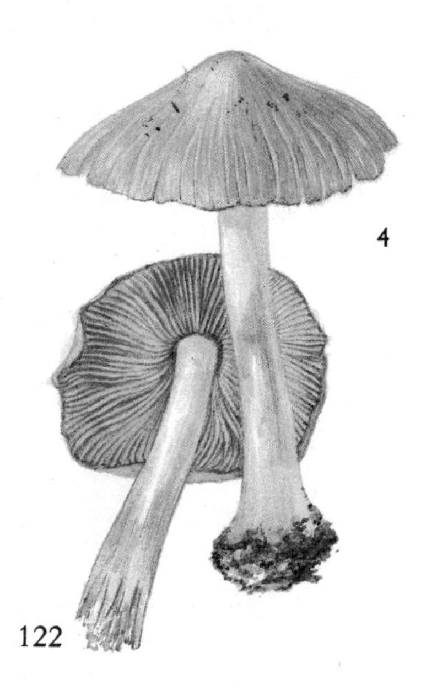


*Inocybe patouillardii* (1) has cap 2.5—8 cm wide, creamy, silky, fibrilous, turning carmine red when touched or when old. Gills whitish, greyish or olive brown, sometimes with red patches. Stipe 5—10/1—2 cm, whitish, creamy, with red patches. Flesh (2) is white, turns red in blotches when old; taste and aroma pleasant. Spore powder ochre brown; spores pale ochre, 9—14/5—8 µm.

*Inocybe patouillardii* is often confused with edible spring mushrooms, such as *Calocybe gambosa* or *Entoloma clypeatum* (3); other *Inocybe* species are confused with the Fairy-ring Champignon.

*Inocybe godeyi* (4) has cap 3—5 cm wide, silky, whitish, turning red when bruised or when old. Gills earth yellow, sometimes with red patches. Stipe

2.5—7/0.6—1 cm, has the same colouring as the cap, with a flat bulb at base. Flesh white, turning red when bruised or old; taste unpleasant, aroma spermatic. Spore powder ochre brown; spores light ochre, 9—12/5—7 µm.

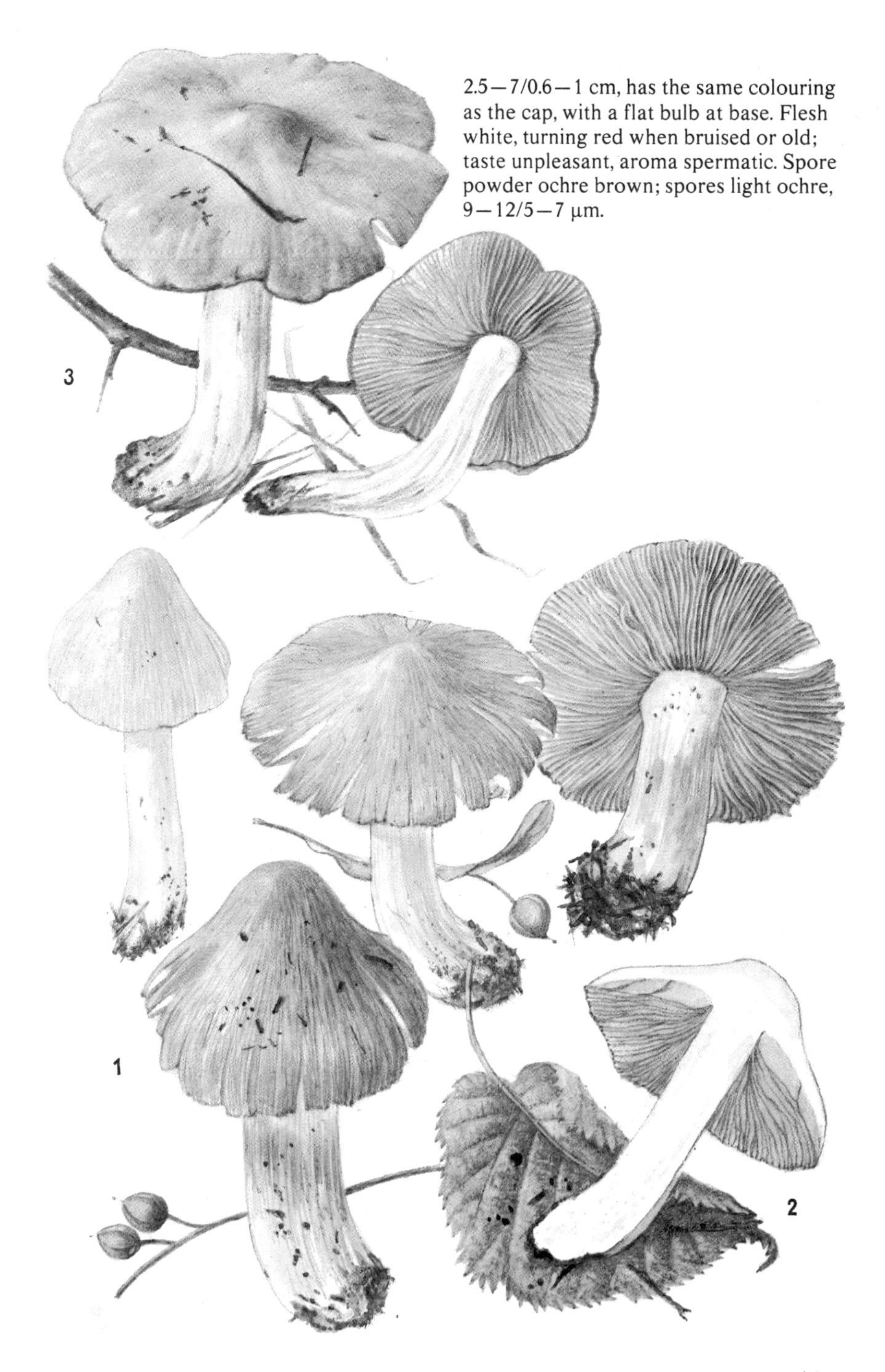

Until 1960 *Cortinarius orellanus* and its many relatives were considered harmless. Then, after frequent cases of poisoning in Poland, some fatal, it was discovered that the cause was *C. orellanus*, despite its pleasant taste and radish-like aroma. Analysis revealed the presence of orellanin, cortinarin, benzoinin and other toxins. From 3 to 24 days elapse before the first symptoms of poisoning become apparent. Then the course of poisoning is very fast; it results in damage to kidneys and death.

*Cortinarius orellanus* is relatively rare; there are still some countries where it has not been discovered. In Europe it grows in autumn and sometimes at the end of summer in deciduous, and rarely in coniferous, forests. It lives in mycorrhizal association with oak and birch and prefers acid soil. It is very difficult to recognize this dangerous mushroom as it resembles many similar species. Classification of *Cortinarius* species poses many problems even for experts. All poisonous species which have been classified belong to the subgenus *Leprocybe*. Mushrooms of this subgroup are characterized by their dry fruit-bodies, caps with ingrown, felt-like silky fibres, slender stipes without swellings and orange to rust-brown gills.

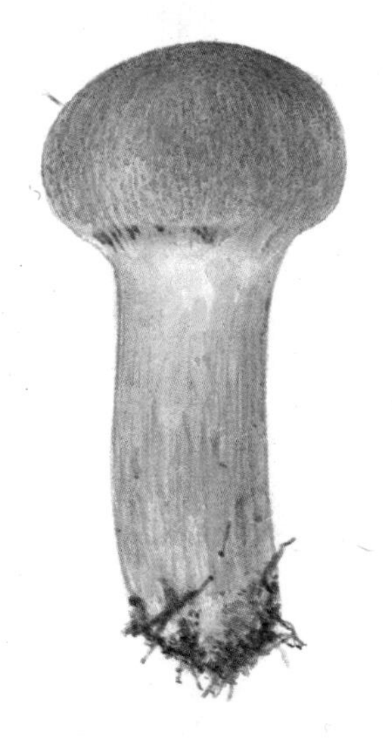

*Cortinarius orellanus* has cap 3—8.5 cm wide, at first subglobose, later flat with an inconspicuous umbo, orange or cinnamon red with a golden tint; dry surface, covered with fine scales. Gills rusty red or cinnamon brown. Stipe 4—9/0.5—1.5 cm, dry, rusty yellow at the base, smooth and with delicate filaments. Flesh is light ochre to rusty yellow (1); taste indistinct, aroma slightly resembles radishes. Spore powder rusty ochre; spores (2) yellow, 8.5—12/5.5—7 µm.

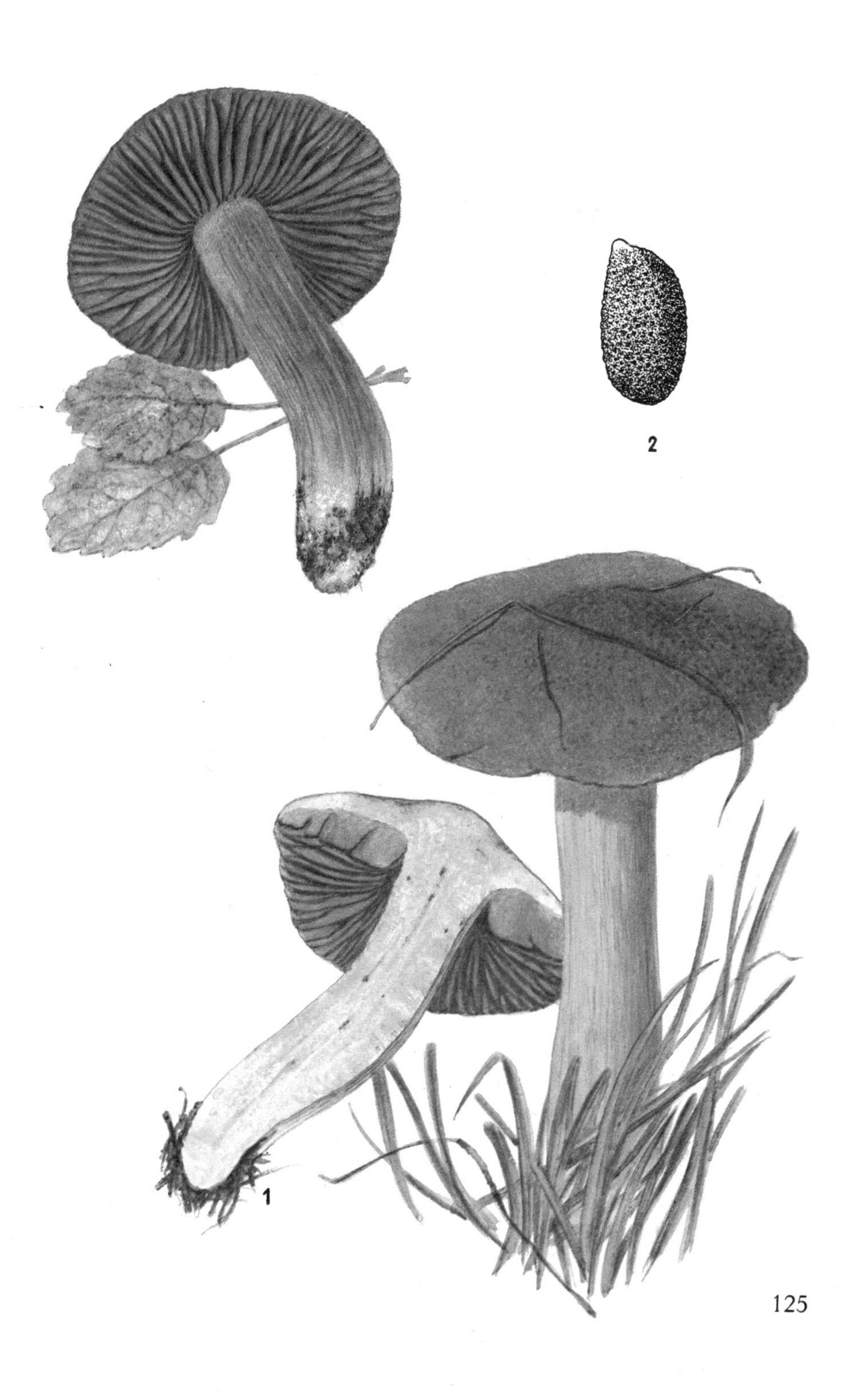
1
2

Its large number of similar species makes the internal classification of the genus *Cortinarius* a difficult task. It is therefore useful to divide this genus into small subgenera and then to attempt the classification of species. The subgenus *Sericeocybe,* which includes *Cortinarius traganus,* has a nonhygrophane cap with a deeply fibrilous and scaly surface and a stipe with a large, tough bulb.

*C. traganus,* one of the most common autumn *Cortinarius* species, is characterized by its purple colouring, which is particularly striking in young fruit-bodies. Old specimens turn brown and so are less conspicuous. *C. traganus* can be found from summer onwards in acid soil in all types of woods especially coniferous forests, and is particularly characteristic of mountainous spruce growths. It is inedible and usually smells of acetylene; the variety *odoratum* is exceptional, as it has a pleasant fruit-like aroma.

*C. traganus* is not the only *Cortinarius* species with purple colour ing and a bulbous stipe. There are at least 5 other similar ones with which *C. traganus* can be confused. The most common is *C. camphoratus,* which has brown-lilac fruit-bodies; the cap has an ochre apex and its surface has deeply ingrown fibrils. The gills are blue or lilac, later even cinnamon yellow. It smells of rotten potatoes. It often shares its habitat with *C. traganus.*

*Cortinarius traganus* has cap 3—10 cm wide, subglobose, later flat and convex, fleshy, surface dry and fibrilous; young fruit-bodies vivid lilac, later pale ochre or leather brown. Gills yellow-ochre, brown to rusty with a purple tint. Stipe 5—10/2—4 cm, lilac, blue-purple, later becoming paler, smooth, swollen at the base into a conspicuous leather brown bulb. Flesh dirty yellow-ochre, unchangeable (1); taste unpleasantly bitter, aroma of acetylene. Spore powder rusty brown; spores (2) brown, with fine warts, 8—10/5—6 μm. This type of spore occurs also in other *Cortinarius* species.

1

## *Cortinarius armillatus* (FR.) FR. *Cortinariaceae*
## Syn.: *Telamonia armillata*

*C. armillatus* is one of the few *Cortinarius* species which is easy to recognize. Belonging to the subgenus *Telamonia*, its fruit-body is dry, without slime; the hygrophane, rust-brown cap is fibrilous and remnants of the veil form strips on the stipe, which is long and enlarged at the base. This species is found in Europe, Asia and North America, most commonly in birch woods or under solitary trees in submountainous regions, and forms mycorrhizal associations with various birch species; it also grows in peat bogs or along their edges, but always close to birch trees. *C. armillatus* needs acid soil and is intolerant of lime. The fruit-bodies develop in groups and can be collected throughout summer until the end of autumn. This mushroom is considered edible in Europe, although of inferior quality; it is advisable to use it in mixtures with other mushrooms. A related species, which is difficult to distinguish from *C. armillatus*, is *C. haematochelis*. However, it has small, almost globose spores as opposed to the larger almond-shaped spores of *C. armillatus*. *C. haematochelis* grows in deciduous and mixed forests and it is also abundant in mountainous spruce stands, for example in the Alps.

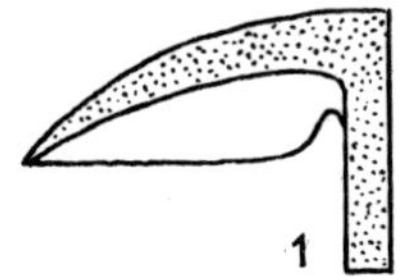


*Cortinarius armillatus* has cap 5—10 cm wide, brick red, pale cinnamon or red rusty brown, fibrilous. Gills sinuate (1), sparse, at first yellowish, later rusty brown with edges in the same colour or lighter. Stipe 6—14/1—3 cm, gradually broadens towards the base, finely filamentous, light brown to light rusty, with one or several red-rusty, slanting strips. Flesh is pale fleshy (2); taste and aroma indistinct. Spore powder pale rusty brown; spores rusty yellow, 7—12/6—7 μm.

2

## Gypsy Mushroom

*Cortinariaceae*

*Rozites caperata* (PERS. ex FR.) KARST.

The Gypsy Mushroom resembles mushrooms of the genus *Cortinarius* and was previously classified with them. Its similarities include the rust-brown spore powder and almond-shaped, warted spores. On the other hand, the Gypsy Mushroom lacks the distinctive cortina between the stipe and the edge of the cap. When its membranous veil breaks, it leaves the ring on the stipe; the lower surface of the ring retains a nearly imperceptible remnant of the veil, called the ocrea.

The Gypsy Mushroom is typical of submountainous and mountainous coniferous forests, where it grows from August to September in acid soil. It can be found close to bilberry growths, under birch trees and often in deciduous forests under beech, probably forming a mycorrhizal association with these plants. The Gypsy Mushroom occurs in Europe, North America and Japan and also far the north in Greenland and Lapland. It can exist to a height of 2,500 m above sea level. The Gypsy Mushroom has an excellent, mintlike flavour, and can be cooked in many ways.

The colour of the Gypsy Mushroom resembles some *Agrocybe* species, namely the edible *A. praecox* and *A. dura*, both of which are abundant in spring and sometimes in summer. They prefer meadows and garden lawns to forests. The *Agrocybe* species have smaller, thinner caps than the Gypsy Mushroom and the stipe is longitudinally streaked and hollow inside. *A. praecox* has a floury taste and aroma, while *A. dura* is bittery in taste.

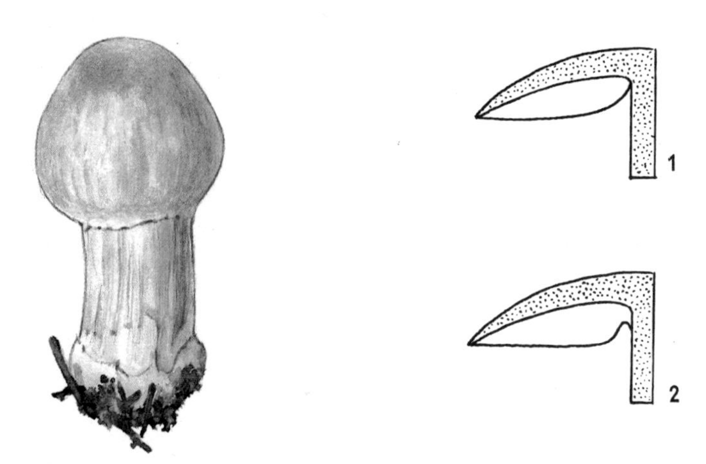

*Rozites caperata* has cap 4—10 cm wide, ovoid to globular when young, later expanded, yellow to ochre; young fruit-bodies have a blue tint and pruinose surface, later become glabrous, cracking and shrivelling in drought. Gills adnexed (1), free or sinuate (2), at first whitish, later earth yellow, with slightly serrated edges. Stipe 5—10/1—2 cm, dirty white with a whitish, membranous ring. Flesh white, unchangeable (3). Taste fungoid, aroma pleasantly spicy. Spore powder rusty brown; spores (4) ochre yellow, 11—14/7—9 µm.

4

## Changing Pholiota

*Cortinariaceae*

*Kuehneromyces mutabilis* (FR.) SING. et SMITH
Syn.: *Pholiota mutabilis*

The Changing Pholiota is a wood mushroom which is greatly sought after for its excellent flavour. In some countries it is cultivated in bulk. It grows in clusters from spring to the end of October, most often on beech, birch, lime, spruce and fir stumps. It is found not only in forests but also in gardens and parks. Changing Pholiota is widely distributed throughout the temperate zone of the northern hemisphere both in lowlands and highlands. It can be cooked in many ways, but the tough stipe is best removed. A rare, related species, *K. vernalis,* has an overall lighter colouring; its stipe is never scaly but filamentous and it grows in spring in the mountains. It is also edible.

The Changing Pholiota is easily confused with deadly poisonous mushrooms of the genus *Galerina,* some of which, such as *G. marginata* and *G. autumnalis,* contain the same toxins as the Death Cap. The first symptoms appear after 10 to 14 hours, when the function of the liver has already been adversely affected. *Galerina marginata* grows on coniferous stumps, especially in montainous areas. However, it differs from the Changing Pholiota in having a smooth stipe, ochre-brown gills and a mealy taste.

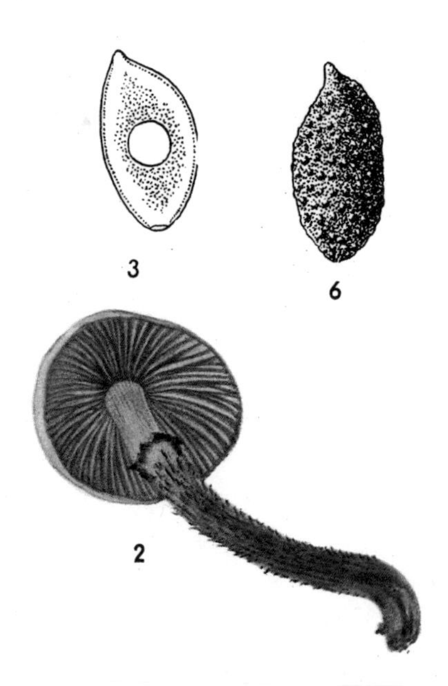


*Kuehneromyces mutabilis* (1) has cap 3—6 cm wide, hygrophane, cinnamon brown, yellow-brown, with darker margins in damp weather, grooved. Gills pale, pale brownish to rusty brown. Stipe is 3—6/0.4—0.8 cm, almost rusty-brown, scaly under the disappearing ring (2). Flesh whitish; taste and aroma meaty and fungoid. Spore powder rusty brown; spores (3) brown, 6—7/3—4.5 μm.

*Galerina marginata* (4) has cap 1.5–4 cm wide, hygrophane, ochre brown, yellow-brown, with a grooved margin. Gills yellowish-brown to ochre brown. Stipe 3–6/0.2–0.4 cm, smooth, ochre to honey or rust-brown, darker at the base (5). Flesh light yellow-brown; taste and aroma mealy. Spore powder rusty brown; spores (6) brown, 8.5–11/5–6.5 µm.

## Buckler Agaric

*Entolomataceae*

*Entoloma clypeatum* (L. ex FR.) KUMM.
Syn.: *Rhodophyllus clypeatus*

Many *Entoloma* species are poisonous. *E. clypeatum* is an exception, and is one of the most popular and tasty mushrooms. One reason for its popularity is that it grows early in spring. From the beginning of April to the end of May it can be found in groups or strips in grass, always outside forests. It likes stands along forest edges, orchards and parks, where it grows under trees and shrubs of the Rosaceae family, for example hawthorn and roses, and also under certain fruit trees with which it forms mycorrhizal associations. *E. clypeatum* is widely distributed throughout the temperate zone of the northern hemisphere. It is closely related to *E. prunuloides*, which is sometimes described as its summer form. *E. prunuloides* has the same habitat and colouring; it is edible and grows only in summer. The two species differ only in microscopic features. *E. clypeatum* is often confused with some poisonous species, for example *E. sinuatum*. Its fruit-bodies are large and fleshy, with a leathery ochre or ash cap. It does not grow in spring, but from June in deciduous forests on lime-rich soil.

Another poisonous relative, *E. vernum*, grows in grass in the same habitat as *E. clypeatum*, also occasionally in coniferous forests. It can be recognized by its smaller size and black-brown, club-shaped cap which has a silky sheen.

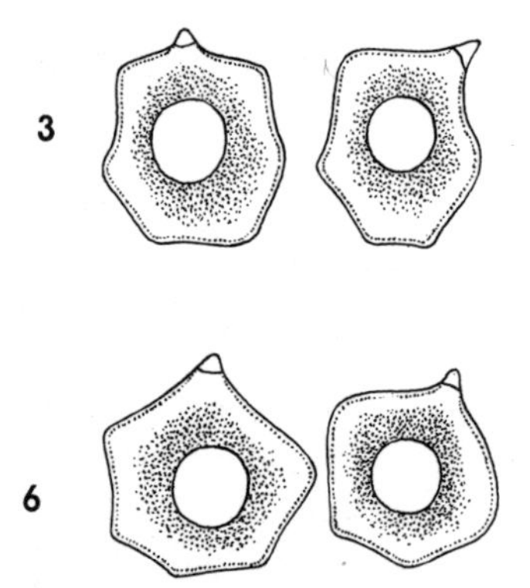


*Entoloma clypeatum* (1) has cap 3—10 cm wide, with an umbo in the centre, expanded, with lobed margin, grey to grey-brown, with silky fibrils. Gills white, later flesh-coloured (2). Stipe 4—15/0.6—2 cm, white, fibrilous. Flesh white; taste and aroma mealy, cucumber-like. Spores (3) pink, 8—11/5—9 μm.

*Entoloma vernum* (4) has cap 2.5—5 cm wide, hygrophane, cone-shaped, broadly campanulate with an umbo, olive brown, grey-brown to black-brown, with a silky sheen. Gills pale grey, later grey-red (5). Stipe 3—8/0.3—0.5 cm, often flattened, longitudinally fibrilous, lighter than the cap. Flesh whitish; taste and aroma indistinct. Spore powder of both species pink; spores (6) light pink, 9—10/7—8 μm.

## Leaden Entoloma

*Entolomataceae*

*Entoloma sinuatum* (BULL. et FR.) KUMM.
Syn.: *Entoloma lividum*

According to statistical data available in some countries, *Entoloma sinuatum* is the main cause of mushroom poisoning. However, this seems improbable as *E. sinuatum* is a rare species in Europe. More likely the poisoning is caused by other *Entoloma* species which are very similar but more common. Poisoning caused by these mushrooms is very severe; general sickness, headache, dizziness and violent vomiting and diarrhoea appear in 1 to 3 hours and last for 3 days. *Entoloma* species are confused with the delicious *Calocybe gambosa, E. clypeatum* and *Lepista nebularis.*

*E. sinuatum* grows from summer to autumn and very rarely in spring in deciduous forests, where it forms mycorrhizal association with beech and oak, preferring soil with a high lime content. The species likes a warm climate and therefore it is rare in northern Europe, while it is common in southern France and other areas of southern Europe; it never grows in mountains. It closely resembles the poisonous *E. rhodopolium,* which has a glabrous, grey or grey-yellow cap with no fibrils, and gills which are white, not yellow, when young. It has a mealy taste and aroma. Another poisonous relative, *E. nidorosum,* has a grey-brown cap, pink-brown gills and a strong odour of nitric acid. Both species grow in deciduous forests.

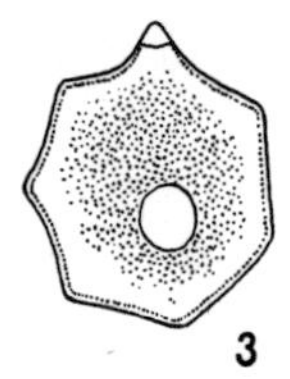


*Entoloma sinuatum* has cap 6—20 cm wide, fleshy, nonhygrophane, long and campanulate when young, later flat and convex, whitish, ash grey to leathery ochre, with a silky sheen. Gills sinuate-decurrent (1), yellowish when young, later reddish. Stipe 6—10/1—2.5 cm, whitish, finely fibrilous, enlarged at the base. Flesh white, filamentous (2); taste unpleasant, slightly floury, aroma floury. Spore powder pink to flesh red; spores (3) are pink, 8—10/7—8.5 μm.

2
1

## Inrolled Paxil, Brown Roll-rim



*Paxillus involutus* (BATSCH) FR.

The Inrolled Paxil and *P. atrotomentosus* can be found in forests during dry seasons when other mushrooms stop growing. The Inrolled Paxil thrives in all types of forests throughout the summer and autumn. Until recently it has been considered edible though of inferior quality. However, after several fatalities it was discovered that repeated consumption of the Inrolled Paxil produces antibodies (agglutinins) to the antigens in the mushroom. These antibodies remain in the body and their amount increases with repeated consumption of the mushroom. When a certain level is reached, they start to act not only against the fungoid antigens, but also against the red cells which they dissolve (hemolyze). Poisoning can therefore appear soon or after a long time, even after several years, depending on the sensitivity of the individual. Symptoms vary, but they usually begin with dizziness and stomachache and end in malfunction of the kidneys.

*P. atrotomentosus* is still considered non-poisonous, but of very inferior quality. Its large, fleshy fruit-bodies, usually with an eccentric stipe, develop on stumps or roots of coniferous trees, particularly on spruce and pine.

*Paxillus involutus* (1) has cap 5—12 cm wide, smooth, ochre brown to red-brown, felt-like along the inrolled margin. Gills shortly decurrent, yellowish, brownish (2,3). Stipe 2—6/1—2 cm, yellowish. Flesh yellowish to brownish; taste and smell slightly acid. Spore powder brown; spores (4) yellow-brown, 8—10/4.5—6 µm.

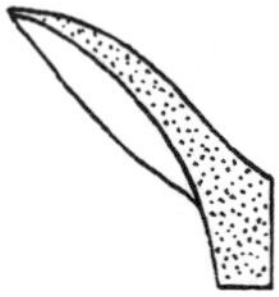

7

3

8

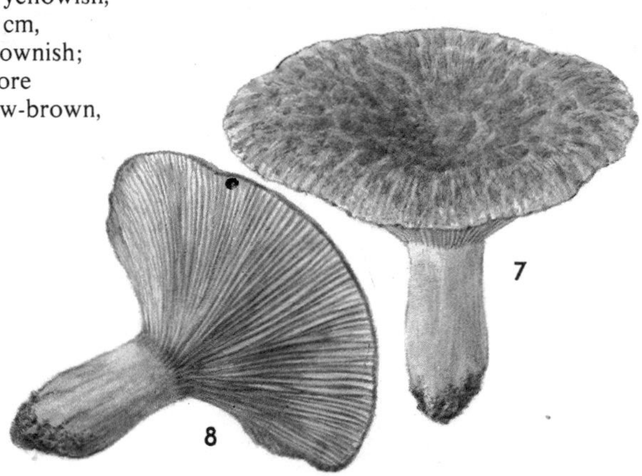

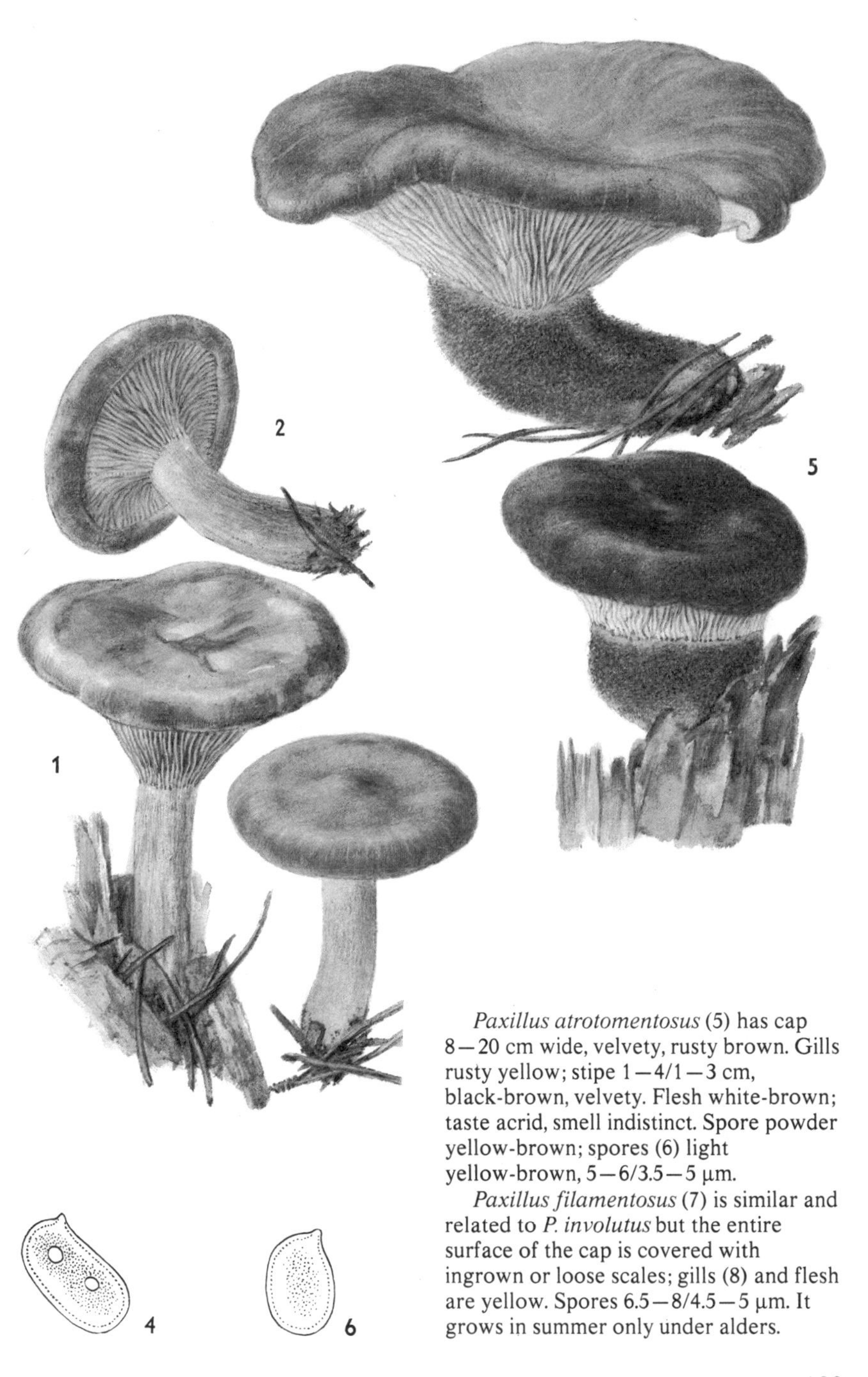


*Paxillus atrotomentosus* (5) has cap 8—20 cm wide, velvety, rusty brown. Gills rusty yellow; stipe 1—4/1—3 cm, black-brown, velvety. Flesh white-brown; taste acrid, smell indistinct. Spore powder yellow-brown; spores (6) light yellow-brown, 5—6/3.5—5 µm.

*Paxillus filamentosus* (7) is similar and related to *P. involutus* but the entire surface of the cap is covered with ingrown or loose scales; gills (8) and flesh are yellow. Spores 6.5—8/4.5—5 µm. It grows in summer only under alders.

# *Chroogomphus rutilus*



(SCHFF. ex. FR.) O. K. MILLER

Syn.: *Gomphidius rutilus*

Mushrooms of the genus *Gomphidius* are mycorrhizal and closely dependent on coniferous trees. In Europe they are represented by a small number of species, but in North America their number is much greater, since there are many more species of conifers. *Chroogomphus rutilus* is widespread throughout Europe, Asia and North America. It has been also introduced to the southern hemisphere together with the pine. It grows gregariously in summer and autumn in coniferous forests only, under certain pine species; its habitat stretches from the lowlands to the subalpine belt, where it grows under dwarf pines. It has a characteristic brown-red cap, which is sticky in damp weather but smooth and glossy in dry conditions.

*Gomphidius glutinosus* has the same area of distribution as *Ch. rutilus.* In Europe it forms mycorrhiza with spruce; it is most abundant in highlands but grows in lowlands as well. It appears from June to October. In North America it also grows under fir; in northern California it occurs as late as December.

All *Gomphidius* species are good edible mushrooms which can be cooked in many ways.

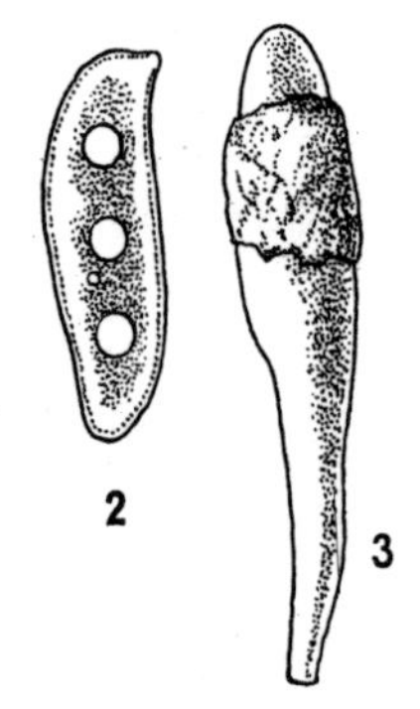


*Chroogomphus rutilus* (1) has cap 3—10 cm wide, red yellow-brown, sticky, with yellow-orange gills. Stipe 5—8/1—1.5 cm, orange yellow-brown. Flesh orange-yellow; a drop of ferrous sulphate colours it rusty. Taste and aroma pleasantly fungoid. Spore powder olive black; spores (2) brown-yellow, 17—22/6—8 µm. Cystidia often incrusted (3).

*Chroogomphus helveticus* (4) has cap with dry surface, felt-like and filamentous. This mountain species forms mycorrhiza with spruce and *Pinus cembra.*

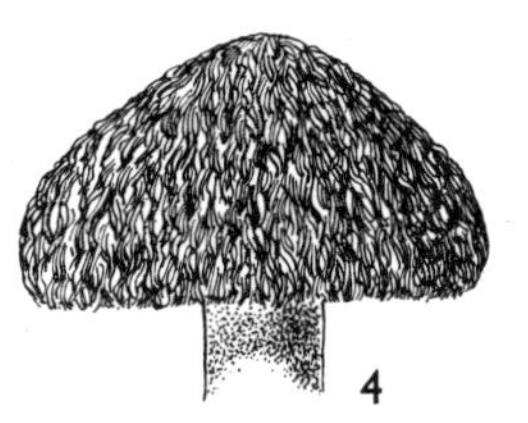


*Gomphidius glutinosus* (5) has cap 5—13 cm wide, grey-brown, slimy. Gills whitish, later becoming black. Stipe 5—10/1.5—2 cm, whitish, lemon yellow at base. Taste and aroma fungoid. Spore powder blackish; spores brown, 18—22/5—7 μm, same shape as those of *Chroogomphus rutilus.*

## Larch Boletus

*Boletaceae*

*Suillus grevillei* (KLOTZSCH) SING.
Syn.: *Boletus elegans*

The Larch Boletus, always found in association with larches, is one of the commonest of *Boletus* species. It has a lemon yellow cap and pores; the cuticle of the cap, as in other *Suillus* species, is slimy and can be peeled off easily. The Larch Boletus can be found from June to November in forests under various species of European, Asian and American larches from lowlands to highlands, growing on needles or in grass, usually in large numbers and always within the root system of the tree. It is quite common under local or exotic larch trees, planted in parks and gardens. The Larch Boletus can be collected from the same spot over decades. Only when its host tree perishes does the mushroom slowly cease to fructify. The Larch Boletus is abundant in Europe, North America, northern Africa, Japan, Australia and New Zealand and has also been introduced to South America. Edible, and with an excellent flavour, it can be cooked in many ways. It is advisable to peel off the slimy cuticle before use.

*S. aeruginascens* also forms mycorrhizal associations with larch trees; it is entirely grey-brown or grey-yellow. It is edible but less common in Europe and North America.

*Suillus grevillei* (1) has cap 4—15 cm wide, subglobose, later flattened, lemon yellow, golden yellow, slimy. Openings of tubes (pores) lemon yellow but long concealed by yellow-white veil. Stipe 4—12/1—2.5 cm, yellow with brown

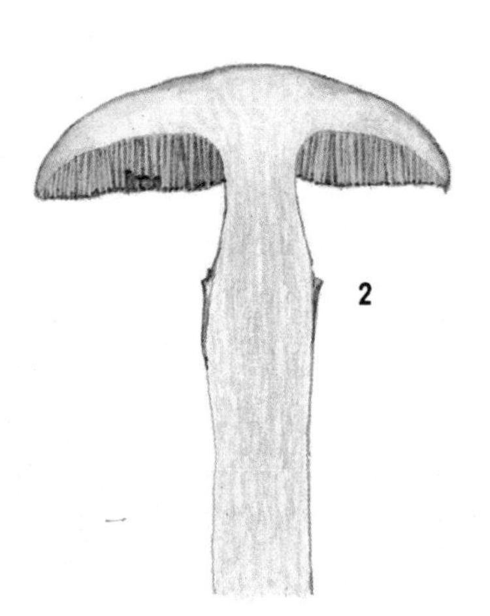


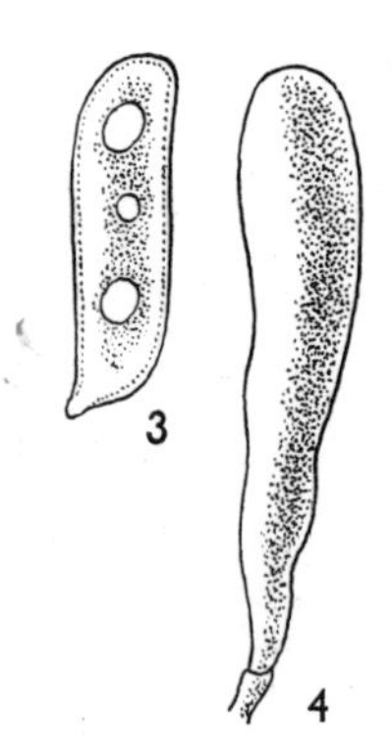

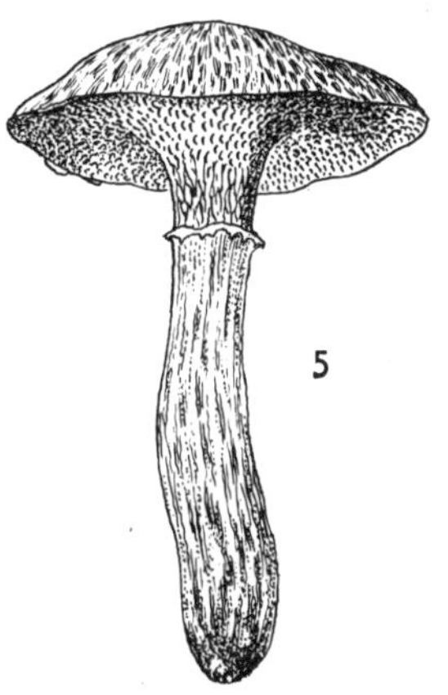


striations and yellowish ring. Flesh yellow (2), soon turns pink-purple; a drop of ferrous sulphate turns it brown-grey. Taste and aroma fungoid. Spore powder olive brown; spores (3) pale yellow, 8–12/3–5 μm. Cystidia yellow, incrusted in brown (4).

Another species growing under larches, *S. tridentinus (5)*, occurs rarely only on lime-rich soil. Recognizable by orange-brown scaly cap and bright red decurrent tubes. Also edible.

## Yellow-brown Boletus, Slippery Jack *Boletaceae*

*Suillus luteus* (L. ex FR.) S. F. GRAY

The Yellow-brown Boletus and the more common Granulated Boletus *(S. granulatus)* form mycorrhizal associations with pines, mostly those species in which the needles are jointed in twos, such as *Pinus sylvestris, P. nigra* and *P. mugo,* which can be found in mountainous and subalpine belts. In the Mediterranean region these boleti can be found under *Pinus halepensis* or *P. pinea.* They were introduced together with pine trees to South America and Australia. They grow from summer to autumn and in southern countries from spring. As far as flavour is concerned, they are among the most delicious mushrooms.

There are no mushroom species in Europe with which the Yellow-brown Boletus could be confused, but in North America there are several closely related, edible species.

The Granulated Boletus has a stipe with a granulated surface and in fresh specimens the pores are covered with droplets of milky liquid. The ring is also missing from the stipes of several other *Suillus* species, for example *S. bovinus,* which differs from other similar species by its ochre-brown pores which are not yellow even in young mushrooms, and *S. placidus,* which has a whitish cap. *S. plorans,* which grows only under *Pinus cembra* high in the mountains, can be recognized by its ochre-brown cap which has deeply ingrown fibrils. Its pores also have milky droplets.

*Suillus luteus* (1) has cap 4—12 cm wide, brown, viscous; pores pale yellow. Stipe 3—11/1—2.5 cm, yellowish, with brown granules above the black-brown ring. Flesh is white or yellowish (2); a drop of ferrous sulphate colours it grey-blue; taste fruity, pleasant aroma. Spore powder brown; spores light yellow, 7—10/3—3.5 μm.

*Suillus granulatus* (3) has cap 3—10 cm wide, brown, yellow-brown, slimy. Pores ochre-yellow, often with milky droplets (4). Stipe 4—5/0.8—2 cm, pale yellow, with yellow to brown-yellow granules in upper section. Flesh is whitish; taste is slightly acidy, aroma fruity. Spore powder ochre to yellow-brown; spores pale yellow, 8—10/2.5—3.5 μm.

1
2
1
3
4

## Variegated Boletus

*Boletaceae*

*Suillus variegatus* (SWARTZ ex FR.) KUNTZE
Syn.: *Boletus variegatus*

The Variegated Boletus thrives in pine habitats, especially in acid soils. It is abundant in lowlands and hills, where it forms mycorrhiza with pine species such as *Pinus sylvestris* and *P. nigra;* it also grows under *P. uncinata* in peat-bogs and high in the mountains in mycorrhizal association with *P. mugo.* It can be found from June to November. In flavour it equals other *Suillus* species. The centre of its distribution is Europe, it has been found as far north as Lapland and as far south as Corsica; its most easterly habitat is the Caucasian mountains. It does not grow in North America or East Asia. The Variegated Boletus can be confused with the edible *S. bovinus,* which also grows under pine trees. However, the fruit-bodies of the latter are completely ochre-brown and the surface of their cap is glabrous, not felt-like.

The Hollow-stemmed Boletus *(Boletinus cavipes)* can be easily recognized by its hollow stipe, brown-red, scaly cap and large tube openings (pores). It grows exclusively under larch throughout the temperate zone of the northern hemisphere, mainly from July to November. It is a good edible mushroom. Larch trees also provide the habitat for the edible *S. tridentinus,* which is distinguished by the small red-orange pores and full rather than hollow stipe.

*Suillus variegatus* (1) has cap 5—14 cm wide, yellow-brown, velvety, slimy when damp. Pores olive-brown (2), turning blue when bruised. Stipe 3—9/2.5—4 cm, yellow-brown. Flesh yellowish, soon turning blue-green; taste pleasant, aroma fruity. Spore powder olive-brown; spores (3) pale yellow, 8—11/3.5—4 μm. Cystidia club-shaped (4).

*Boletinus cavipes* (5) has cap 4—12 cm wide, reddish brown, scaly. Pores large, yellow or green-yellow (6), covered with a whitish veil when young. Stipe 3—9/0.8—3.5 cm, brownish, hollow, with a brown ring. Flesh is yellowish, unchangeable (7). Taste mild, sometimes acrid, aroma pleasant. Spore powder light yellow-green; spores (8) light yellow, 7—10/3—4 μm. Cystidia are bottle-shaped (9).

1
2
3
4
5
6
7
8
9

## Red-cracked Boletus

*Boletaceae*

*Xerocomus chrysenteron* (BULL. ex ST. AM.) QUÉL.
Syn.: *Boletus chrysenteron*

*Xerocomus chrysenteron* is a mycorrhizal mushroom, dependent on various species of deciduous and coniferous trees. It is abundant nearly everywhere from summer to autumn. A good edible mushroom, it is particularly tasty when young, but old specimens quickly become soft and are usually attacked by insect larvae. During wet weather, *X. chrysenteron* is sometimes attacked by an *Ascomycetes* fungus, *Apiocrea chrysosperma;* its white mycelium with yellow spores cover the entire fruit-body.

*X. chrysenteron* can be easily recognized by the reddish base of its stipe and the regularly cracked cap of the adult fruit-bodies. *X. truncatus* grows in similar habitats as *X. chrysenteron;* it is identical in appearence and can be distinguished only by microscopic examination of the spores. The same applies to *Boletellus intermedius* and *B. zelleri.* All four species are found in Europe and North America; the former two in deciduous forests and the latter two in coniferous forests.

The Yellow-crack Boletus *(X. subtomentosus)* also resembles *X. chrysenteron* and it grows abundantly in the same habitats at the same time. It has yellowish pores and a yellow stipe. If the cap is cracked on the surface, its cracks are never red. The Yellow-crack Boletus can be confused with the rare *X. spadiceus,* which has a smooth, brown-red to dark brown cap and a delicate red network on its yellow stipe; it grows in coniferous forests and is edible.

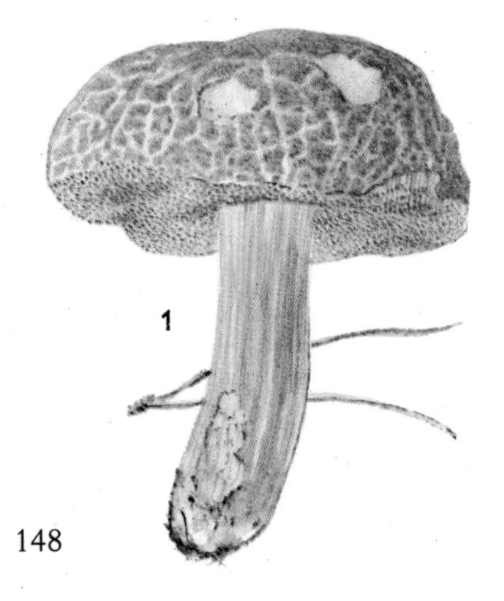


*Xerocomus chrysenteron* (1) has cap 3—8 cm wide, brown, regularly cracked when adult, cracks red inside. Pores yellow-greenish, turning blue-green when bruised. Stipe 3—6/0.7—1.4 cm, yellow, carmine red in lower half. Flesh yellow, slowly becoming blue (2). Spores (3) light ochre-brown, spindle-shaped, 13—15/5—6 μm.

Spores of *Xerocomus truncatus* are suddenly cut off at apex (4).

Spores of *Boletellus intermedius* and *B. zelleri* have longitudinal ribs (5).

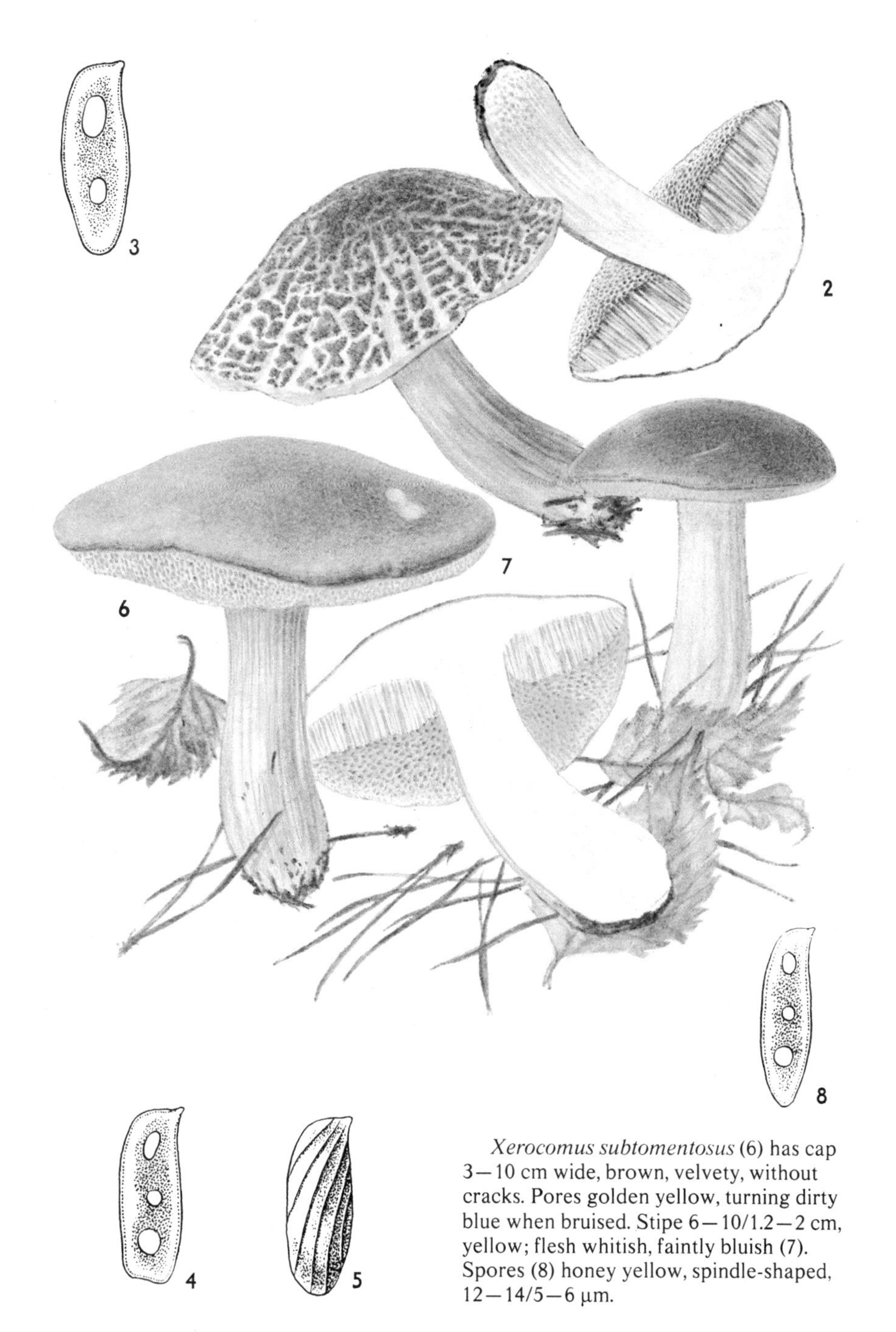


*Xerocomus subtomentosus* (6) has cap 3–10 cm wide, brown, velvety, without cracks. Pores golden yellow, turning dirty blue when bruised. Stipe 6–10/1.2–2 cm, yellow; flesh whitish, faintly bluish (7). Spores (8) honey yellow, spindle-shaped, 12–14/5–6 μm.

## Cèpe, Edible Boletus, Penny-bun Bolete

*Boletaceae*

*Boletus edulis* BULL. ex FR.
Syn.: *Boletus bulbosus*

Together with *B. aestivalis* and *B. pinicola* the Edible Boletus is considered the best flavoured mushroom. Its mycorrhizal partner is spruce and therefore it can be found both in natural and cultivated coniferous forests; it accompanies the spruce to the upper limits of its occurrence. The Edible Boletus also grows under solitary spruces in gardens and parks. Occasionally it can also be found under other coniferous species. The Edible Boletus is widespread throughout the temperate zone of the northern hemisphere, but its distribution varies from one area to another; it is more common in Europe than in North America.

Characteristic features of the Edible Boletus are its light chestnut brown cap, pores which are white when young and yellow-green later and its whitish stipe with white reticulations (network) in the upper section. This mushroom is often confused with *B. aestivalis and B. pinicola. B. aestivalis* has a leathery brown cap and a brown stipe, which is entirely covered with distinct white reticulations. It grows in deciduous forests under oak and beech trees. *B. pinicola* has an almost red-brown cap and the greatly enlarged stipe is entirely covered with red-brown reticulations. It can be found under pine.

The Bay Boletus *(Xerocomus badius)* does not equal the Edible Boletus in flavour; nevertheless it is highly valued. It grows from June to November in coniferous forests under various types of pine and spruce trees, and in European forests also under beech and oak trees. It prefers acid soil found in lowlands and high in the mountains. It is widespread throughout in whole temperate zone of the northern hemisphere, particularly in Europe.

*Boletus edulis* (1) has cap 6—25 cm wide, pale chestnut brown. Pores white when young, later yellow-green. Stipe 2.5—20/1.5—7 cm, whitish, brownish when old, frequently covered with white reticulations (network) in the upper section. Flesh white; taste and aroma appetizingly fungoid. Spore powder olive-brown; spores (2) honey yellow, 14—17/4.5—5.5 µm.

*Xerocomus badius* (3) has cap 4—15 cm wide, chestnut brown, sticky. Pores yellow-green, turning blue when bruised. Stipe 4—12/0.8—4 cm, flesh whitish, becoming faintly blue; taste and aroma fungoid. Spore powder olive; spores (4) yellow, 12—16/4—6 µm.

## *Boletus aestivalis* PAULET ex FR. *Boletaceae*

Syn.: *Boletus edulis* ssp. *reticulatus*
*Boletus atkinsonii*

*Boletus aestivalis* forms mycorrhizal associations with oaks and also beeches and lime trees; it grows in deciduous forests and in parks from spring to autumn. This excellent mushroom is very common throughout Europe, especially in the south; it probably also grows in North America and northern Africa. It is characterized by its leathery brown cap, yellow-green pores and light brown stipe, which is covered along its entire length by distinct, white reticulations. *B. aestivalis* is often confused with the Edible Boletus *(B. edulis)* which has a stipe with white reticulations only along its upper section and is associated with spruce. The two species also differ in the anatomical structure of the cap's cuticle. Inexperienced mushroom pickers can also confuse *B. aestivalis* with *Tylopilus felleus,* which is very bitter in taste. However, the latter grows mainly in coniferous forests; it has pink pores and distinct coarse reticulations all over its brownish stipe.

*Boletus pinicola* is considered by some mushroom pickers to be the best-flavoured of all mushrooms. It is very popular in Europe, North America and elsewhere. It grows in summer and autumn in pine forests, mostly under *Pinus sylvestris.* Occasionally it can be found also under spruce, fir and beech trees. The similar *B. aereus* has a chocolate brown cap and slightly paler stipe with brown reticulations. It likes a warm climate and therefore it thrives in oak woods in southern Europe. It too has an excellent flavour.

*Boletus aestivalis* (1) has cap 6—20 cm wide, leathery brown; pores white in young mushrooms, later yellow-green. Stipe 10—25/2—7 cm, light brown, covered with distinct white reticulations over the entire length. Flesh white, unchangeable; taste and aroma pleasantly fungoid. Spore powder olive-brown, spores honey-yellow, 13—20/3.5—6 μm.

*Boletus pinicola* (2) has cap 6—25 cm wide, ridged, dark chestnut brown, pores white when young, later yellow-green. Stipe 6—15/3—10 cm, light red-brown, entirely covered with red-brown reticulations. Flesh white, unchangeable; taste and aroma pleasantly fungoid. Spore powder olive-brown; spores brownish, 12—17/4—5 μm.

1
2

## Scarlet-stemmed Boletus

*Boletaceae*

*Boletus calopus* FR.

The Scarlet-stemmed Boletus is often confused with the Satan's Boletus *(B. satanas).* The two species resemble each other in the colour of the cap and stipe base, and in the flesh which turns blue when cut. However, they are distinguished by the colour of their pores; in the Satan's Boletus they are red, while in the Scarlet-stemmed Boletus they are yellow. The Scarlet-stemmed Boletus is inedible because of its bitter taste, but the Satan's Boletus, though only edible when cooked, has a pleasant taste. The Scarlet-stemmed Boletus prefers submountainous and mountainous coniferous forests, while the Satan's Boletus grows only in deciduous forests in low-lying, warm areas in soil with a high lime content. The Satan's Boletus is therefore rare, while the Scarlet-stemmed Boletus is quite common, particularly in autumn, in all European mountains. However, its occurrence outside Europe is sporadic.

*B. appendiculatus* prefers a warm climate and therefore its numbers increase towards southern Europe. It also occurs in northern Africa and in Asia. It grows in mycorrhizal association with oak and beech on lime-rich soils in warm, low-lying deciduous forests. It is characterized by a deep-rooting stipe, a quarter of which is hidden in the ground; and by its brown cap, flesh that turns slightly blue and the delicate, lemon-yellow reticulations on the stipe. It has an excellent taste but it can be confused with the inedible *B. impolitus,* which grows in the same habitats. The latter can be recognized by its pale to light brown cap, pale yellow flesh which does not change its colour when exposed to the air, and its light yellow, unreticulated stipe. It smells unpleasantly of carbolic acid.

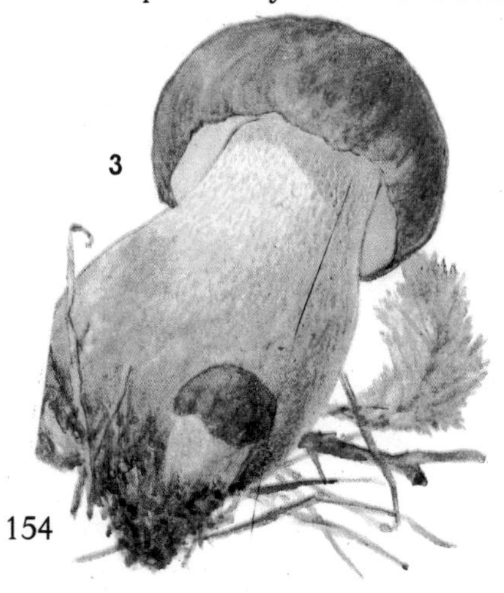


*Boletus calopus* (1) has cap 5—15 cm wide, grey, grey-brown, velvety. Pores turn blue and green when bruised. Stipe 3—15/1—4.5 cm, yellow with yellow reticulations, red with red reticulations at the base. Flesh yellowish, later blue-green, soon becoming pale (2). Taste bitter, aroma pleasant. Spore powder pale ochre-olive; spores pale yellow, 10—14/4—6 μm.

*Boletus appendiculatus* (3) has cap 7—20 cm wide, brown to dark brown. Pores are lemon yellow, golden brown and turn blue-green when bruised. Stipe 5—15/1.5—5 cm, covered with delicate yellow reticulations, turns blue when bruised. Flavour pleasantly fungoid. Spore powder is light olive brown, spores are pale flesh yellow, 10—16/4—6 µm.

*Boletus erythropus* belongs to a group of boleti in which the flesh turns blue when cut and exposed to the air. Because of this, and because its pores become red, *B. erythropus* is often confused with the poisonous Satan's Boletus *(B. satanas)*. The flesh of the Satan's Boletus also turns blue, but the change is very gradual and faint. Other distinguishing features of Satan's Boletus include the grey cap and the net on the purple stipe. The two species also differ in their ecological patterns. *B. erythropus* grows in summer and autumn in coniferous and deciduous forests. In low-lying regions it grows under beech and oak trees; but it is more common in submountainous and mountainous forests, where it forms mycorrhizal association with spruce and fir. It is very rare in lime-rich soil and does not grow in peat-bogs. It is abundant throughout Europe and it also occurs in Africa, Asia Minor and North America. It is edible as long as it is well cooked; otherwise it can cause stomach and intestinal disorder. *B. erythropus* can be confused with *B. luridus*, the flesh of which also turns blue when cut. But the change is less apparent; the cap is lighter in colour and the stipe has no red scales, but distinct red net. Another related species, rare and thermophilous *B. queletii*, has a red-tinted brown cap; the surface of the stipe is granulated, but less distinctly than in *B. erythropus*. The flesh turns blue slowly and less obviously. The stipe is red. *B. queletii* is edible if cooked thoroughly.

*Boletus erythropus* has cap 5—20 cm wide, dark brown or black-brown, finely velvety, turning dark when bruised. Pores yellow when young, later carmine red, turning blue when bruised. Stipe 5—15/2—4 cm, with red scales or granules (1), olive at base, turning blue when bruised. Flesh lemon yellow, turns immediately dark blue when cut (2); taste and aroma fungoid. Spore powder olive-brown; spores olive-yellow, 11—19/5—7 μm.

1
2

## **Lurid Boletus** *Boletaceae*
*Boletus luridus* SCHFF. ex FR.

The Lurid Boletus is one of the earliest spring boleti, appearing as early as May and continuing to develop until October. It is most common in deciduous forests where it forms mycorrhizal associations with oak, beech, lime and birch. It grows also in parks and gardens. The Lurid Boletus likes warm weather in lowlands and uplands. It is relatively rare in submountainous and mountainous areas, where it is replaced by a related species, *B. erythropus.*

The Lurid Boletus is widespread throughout the temperate zone of the northern hemisphere; it is most common in Europe, particularly in the south. In North America it is considered poisonous, but in Europe it is considered poisonous only when raw, and edible when well cooked. Since insufficiently cooked specimens can cause vomiting and diarrhoea, the Lurid Boletus is probably best avoided in the kitchen. It is also possible that it may cause poisoning when alcohol is consumed after it, as with the Common Ink Cap.

The Lurid Boletus closely resembles *B. erythropus.* The main difference between the two species is the surface of the stipe, which in *B. erythropus* is red and scaly or granulated, while in the Lurid Boletus it has a red net. The Lurid Boletus can be easily confused with the relatively rare, thermophilous *B. queletii* which, however, never has a net on the stipe.

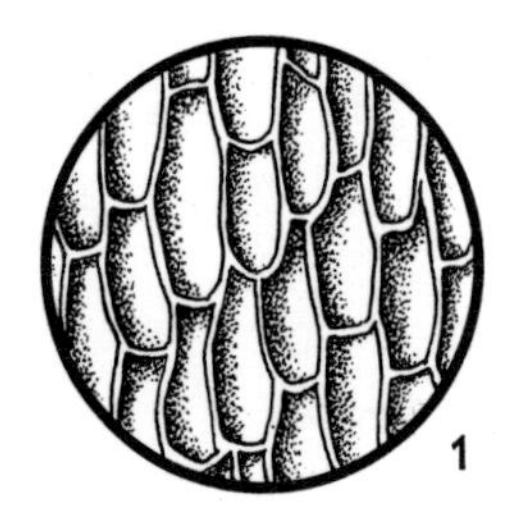


*Boletus luridus* has cap 4—20 cm wide, olive-brown, yellow-brown, light chestnut brown, finely pubescent on the surface, turning dark when bruised. Pores yellow when young, later orange, brick red, turning blue when bruised. Stipe 4—15/2—8 cm, yellow, vine red at base, entirely covered with red net (1), turning blue when bruised. Flesh pale yellow, red at the base of stipe, turning slightly blue (2). Taste and aroma pleasant. Spore powder olive; spores honey yellow, 9—17/5—7 μm.

2

## Satan's Boletus, Satanic Boletus

*Boletaceae*

*Boletus satanas* LENZ

The rare Satan's Boletus is a strongly poisonous mushroom, but only when it is raw, because its toxins decompose with boiling and cease to be dangerous; the cooked mushroom can be therefore eaten without danger. However, even a small piece of the raw mushroom soon causes vomiting and diarrhoea which last several hours. The Satan's Boletus is easily recognized by its dirty grey cap, carmine red pores and bulbous stipe, which is carmine red in the middle and yellow at the top under the cap and at the base. The flesh when cut is white and slowly turns blue. The Satan's Boletus needs lime-rich soil and warm climate, and forms mycorrhiza with beech, oak, hazelnut, chestnut and lime. It appears in small groups from July to September, particularly in southern Europe. It is very rare in North America, where it grows only under oak along the west coast. A similar, also poisonous species, *B. eastwoodiae,* is more common. Other boleti which turn blue and have red pores and therefore may be confused with the Satan's Boletus are *B. erythropus* and *B. luridus.*

*B. radicans* is also often confused with the Satan's Boletus; it grows in the same types of habitat and together with it. The colour of the cap is very similar, but the yellow pores and bitter taste of *B. radicans* are enough to distinguish it. *B. radicans* is a rare inhabitant of Europe, northern Africa and North America. It is inedible.

*Boletus satanas* (1) has cap 10—20 cm wide, dirty grey. Pores yellowish when young, later carmine red (2), turning blue-green when bruised. Stipe 4—12/5—10 cm, swollen, carmine red in the middle, otherwise yellowish, reticulated. Flesh whitish, gradually turns slightly blue (3); taste and aroma slightly acid. Spores yellowish, 10—16/5—7 μm. Distribution and ecology similar to *B. rhodoxanthus* (4), also poisonous when raw; distinguishable by its reddish, grey-brown cap, purple pores and yellow stipe, which has a dense purple net on the surface.

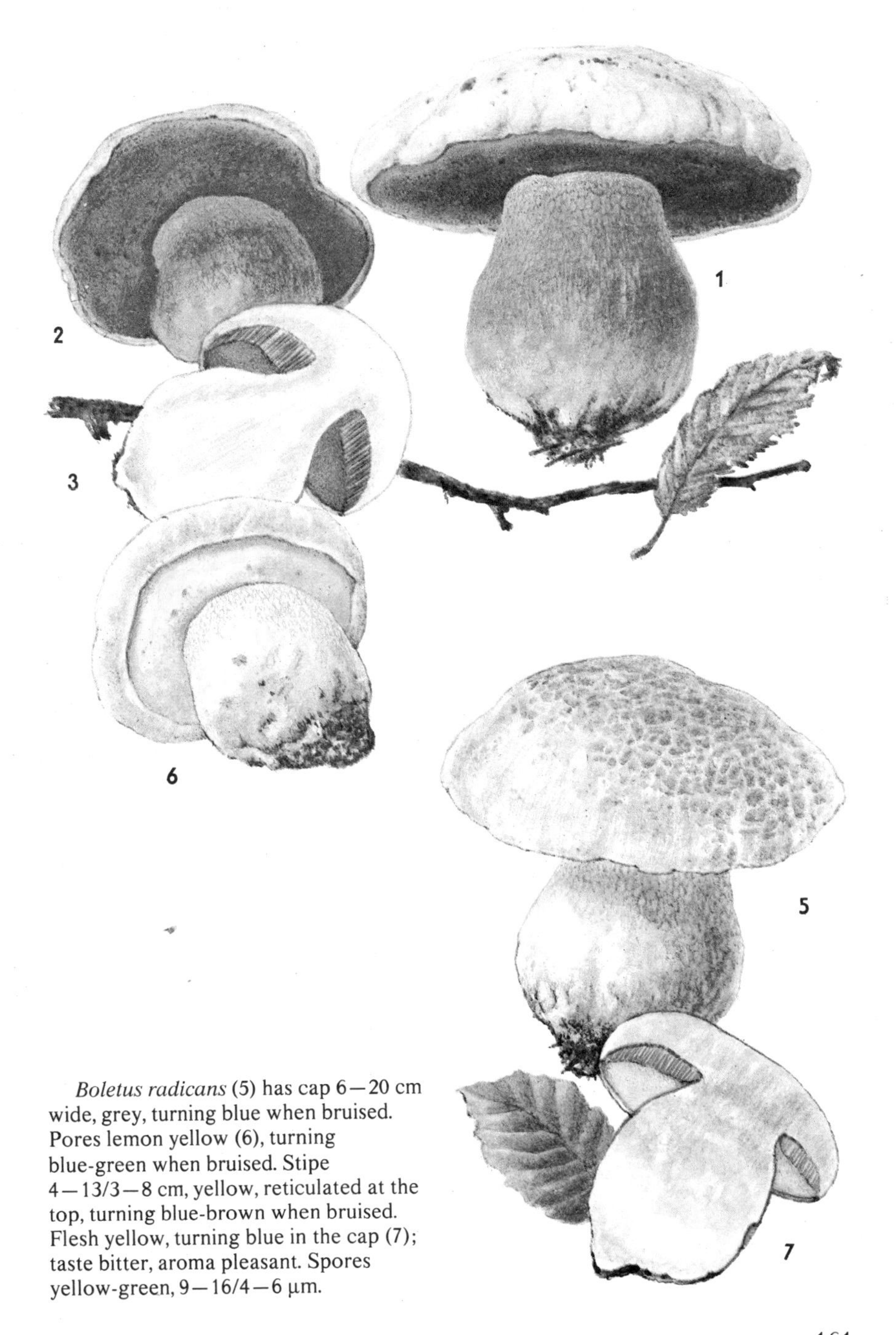


*Boletus radicans* (5) has cap 6—20 cm wide, grey, turning blue when bruised. Pores lemon yellow (6), turning blue-green when bruised. Stipe 4—13/3—8 cm, yellow, reticulated at the top, turning blue-brown when bruised. Flesh yellow, turning blue in the cap (7); taste bitter, aroma pleasant. Spores yellow-green, 9—16/4—6 μm.

## Shaggy Boletus

*Boletaceae*

*Leccinum scabrum* (BULL. ex FR.) S. F. GRAY

*Leccinum scabrum* depends on birch trees for its existence, forming mycorrhizal associations with *Betula pendula* and *B. pubescens* and also with some exotic species planted in parks and gardens. It is abundant from lowlands high up in the mountains, growing from summer to autumn. *L. scabrum* is widespread in the whole temperate zone of the northern hemisphere, being particularly prolific in Europe and North America; it even occurs above the arctic circle in Alaska and Greenland, where it forms mycorrhiza with dwarf birches; it has been introduced to South America.

*L. scabrum* is very rare in acid soil or in peat-bogs, even if birch trees grow in them. In such habitats it is replaced by other related species, such as *L. oxydabile, L. holopus* and *L. rotundifoliae. L. oxydabile* has a light grey-brown cap, and a white stipe with grey or black scales; its flesh turns slightly red when cut and faint yellow at the base of the stipe. *L. holopus* has a whitish cap with blue to grey-green tones, a grey-white stipe and white flesh that becomes blue-green at the base of the stipe. *L. rotundifoliae* grows in mountainous peat-bogs and together with Scandinavian vegetation. Its cap is only 2—7 cm wide, and is coloured ochre grey or pale brownish; the stipe bears white, later brown scales. *L. griseum* grows under hornbeam in lowlands and uplands and is commonly confused with *L. scabrum,* which it resembles very closely in shape and colouring, differing only in its shrivelled, black-brown cap, its network of ribs on the black-brown, scaly stipe, and by the flesh which turns purple-black when cut. All *Leccinum* species are edible and have an excellent flavour.

*Leccinum scabrum* has cap 5—13 cm wide, semiglobular, grey-brown, blackbrown or red-brown, glabrous. Pores white at first, later dirty white and grey. Stipe 8—15/1—3 cm, whitish, entirely covered with dark brown to black scales. Flesh whitish, grey when old, does not change colour when cut (1); a drop of ferrous sulphate colours it blue-grey. Taste and aroma pleasantly fungoid. Spore powder olive brown; spores yellow-brown, 13—20/5—6 μm.

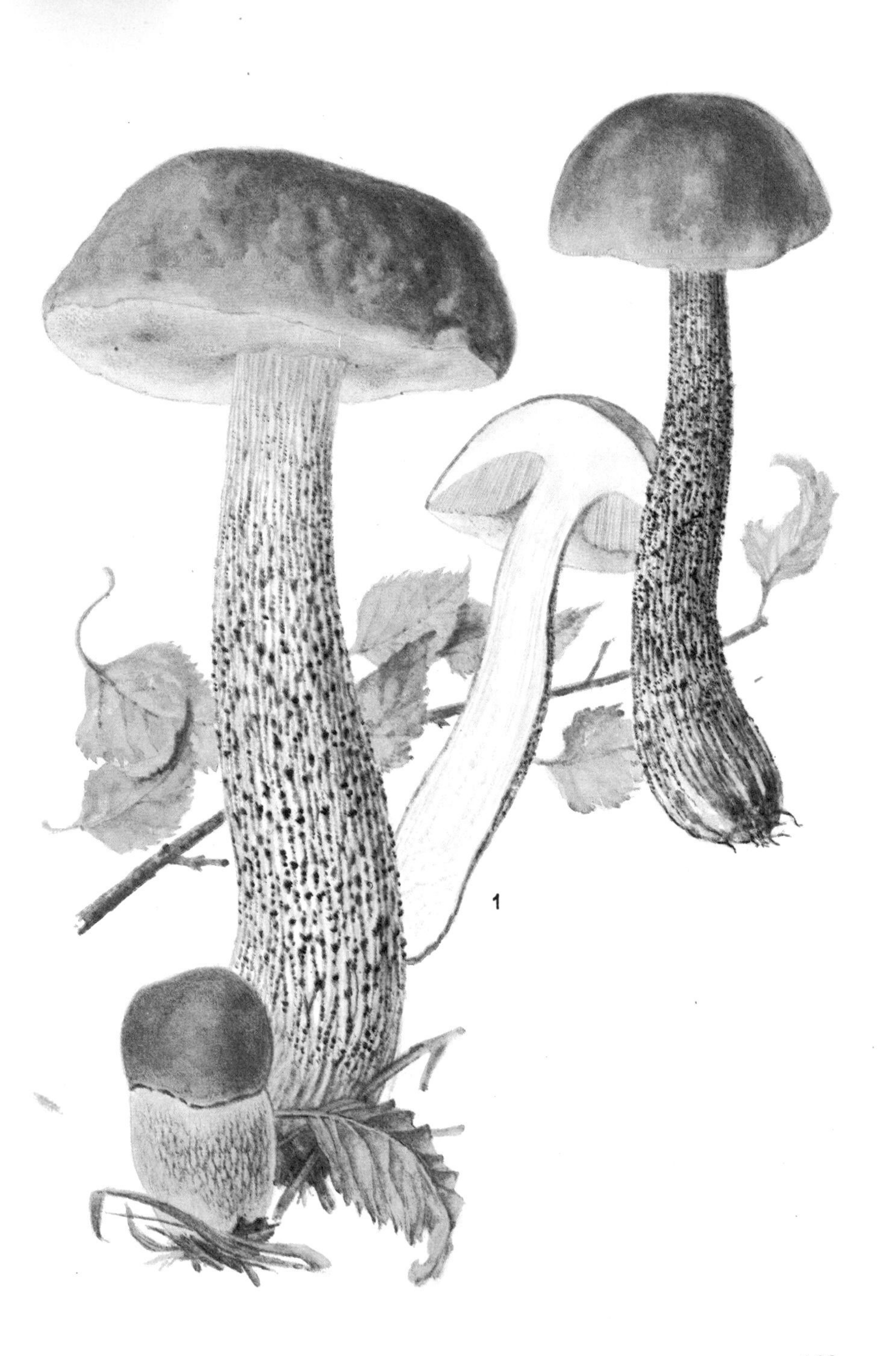
1

## Orange-cap Boletus

*Boletaceae*

*Leccinum aurantiacum* (BULL. ex ST. AM.) S. F. GRAY
Syn.: *Boletus aurantiacus*

*Leccinum aurantiacum* is among the most handsome and delicious of edible mushrooms, and it cannot be confused with any poisonous species. It is characterized by its orange-red cap and whitish stipe covered with brown-red scales. The flesh turns lilac to grey-red when cut. *L. aurantiacum* forms mycorrhizal associations with trees of the genus *Populus.* Most often it grows under the aspen *(P. tremula),* inhabiting light woodland, thickets and parks, occurring even high in mountains. Widespread throughout the whole temperate zone of the northern hemisphere, wherever poplars grow, *L. aurantiacum* is most abundant in central and northern Europe and in North America.

*L. aurantiacum* resembles *L. testaceoscabrum,* which grows only under birch trees; the latter has an orange-yellow and yellow-brown cap. Its pores are already greyish when young; the scales on the stipe are black-brown, and white flesh, when cut, turns pink lilac. The base of the stipe is bluish or greenish. A rare relative, *L. vulpinum,* has a rusty brown cap and its stipe has red-brown to brown scales. The flesh changes colour to lilac and later to brown; the stipe is blue-green at the base. This species forms mycorrhiza with pine species, whose needles are paired, such as *Pinus sylvestris, P. nigra and P. mugo.*

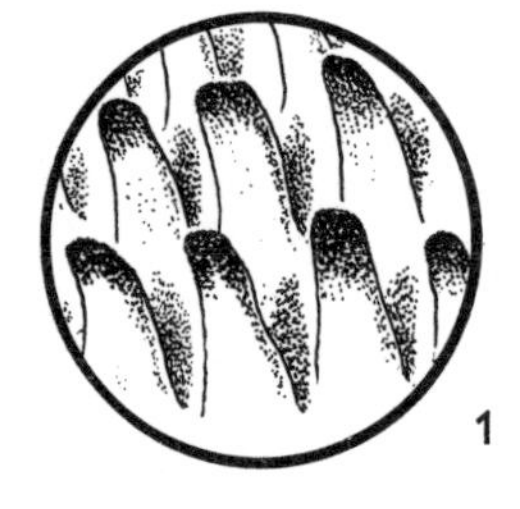


*Leccinum aurantiacum* has semiglobular cap 4—20 cm wide, smooth, orange-red, orange-brown or brown-red. Pores remain whitish for a long time. Stipe 5—15/1.5—5 cm, whitish, often with blue-green patches at the base, with dense covering of coarse, brown-red scales (1). Flesh white; when cut turns lilac to grey-black, sometimes faint green-blue at the base of the stipe (2). Taste and aroma pleasantly fungoid. Spore powder ochre-brownish; spores yellow-brown, 13—17/4—5 μm.

2

## *Leccinum testaceoscabrum* (SECR.) SING. — *Boletaceae*
Syn.: *Boletus versipellis*

*Leccinum testaceoscabrum,* like *L. scabrum,* forms mycorrhiza with birch trees. Both species have almost the same area of distribution, except that *L. testaceoscabrum* also grows in peat-bogs if birch trees are present. In summer and autumn, it grows individually or in groups in forests, gardens and parks under solitary birch trees, it has even been found in the Caucasus at 2,400 m above sea level. In northern tundra it lives in mycorrhizal association with the dwarf birch species *Betula nana* and *B. rotundifolia,* and occurs above the arctic circle in Greenland, Lapland and on Spitsbergen. This excellent edible mushroom can be cooked in many ways.

*L. testaceoscabrum* and *L. aurantiacum* used not to be considered as separate species, although the former grows only under birch trees and the latter only under poplar trees. The two species differ in their colour: *L. testaceoscabrum* has a yellow-brown cap whereas the cap of *L. aurantiacum* is orange-red or orange-brown. The pores in *L. testaceoscabrum* are grey when young and later grey-brown, but in *L. aurantiacum* they remain whitish for a long time. The stipe of *L. testaceoscabrum* is covered with fine, brown-black scales, while the scales of *L. aurantiacum* are brown-red and coarser. The flesh of *L. testaceoscabrum* turns pink-lilac when cut, and is always blue-green at the base of the stipe, while in *L. aurantiacum* it turns lilac and the base of the stipe becomes faintly green-blue only sometimes.

*Leccinum testaceoscabrum* has wide, semiglobular cap 4—15 cm across, delicately tomentose, often with a membranous border along the margin, yellow-orange, yellow-brown or orange-brown. Pores very small and greyish when young, later grey-brown. Stipe 5—10/2—5 cm, whitish, entirely covered with brown-black to blackish scales (1). Flesh white, turns pink-lilac when cut; at the base of stipe almost blue-green (2). Taste and aroma typically fungoid. Spore powder yellow-brown; spores honey brown, 13—16/4—5 μm.

2

# Bitter Boletus

*Boletaceae*

*Tylopilus felleus* (BULL. ex FR.) KARST.

*Tylopilus felleus* is not poisonous, but its bitter, gall-like taste makes it inedible. Its attractive fruit-bodies, which are rarely attacked by insects, grow from moss and are easily mistaken for the Edible Boletus, particularly in their early stages. The two differ, however, first and foremost in their taste; also the rough reticulations (network) on the stipe of *Tylopilus felleus* are brown on a white background, while the delicate net on the stipe of the Edible Boletus is white against a brown background. The pores of *T. felleus* are pink, but in the Edible Boletus they are yellow-green. *T. felleus* grows in coniferous forests from summer to the end of autumn, often appearing in large numbers in acid, sandy soil under pine, in sandy, clay soil under spruce and larch and also occasionally on fallen rotten tree trunks. It is most abundant in lowlands and on hills; it is less common in submountainous regions and it does not occur high in the mountains. Besides the temperate zone of the northern hemisphere *T. felleus* is also found in Australia, New Zealand and South America. It often occurs close to pine trees, with which it probably forms mycorrhizal associations. The genus *Tylopilus* is represented by only one species in Europe, while it has about 10 species in North America. The closely-related North American species *T. plumbeoviolaceus* has a distinct net at the top of the stipe only, and both the cap and stipe are coloured purple. It also grows in coniferous forests and tastes bitter.

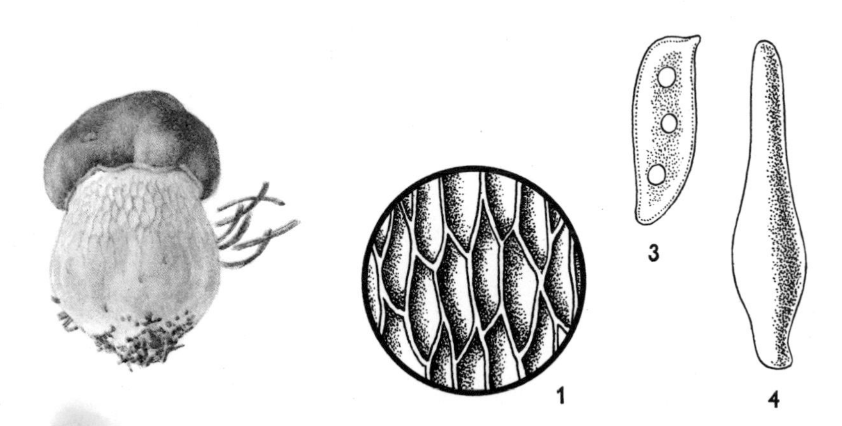

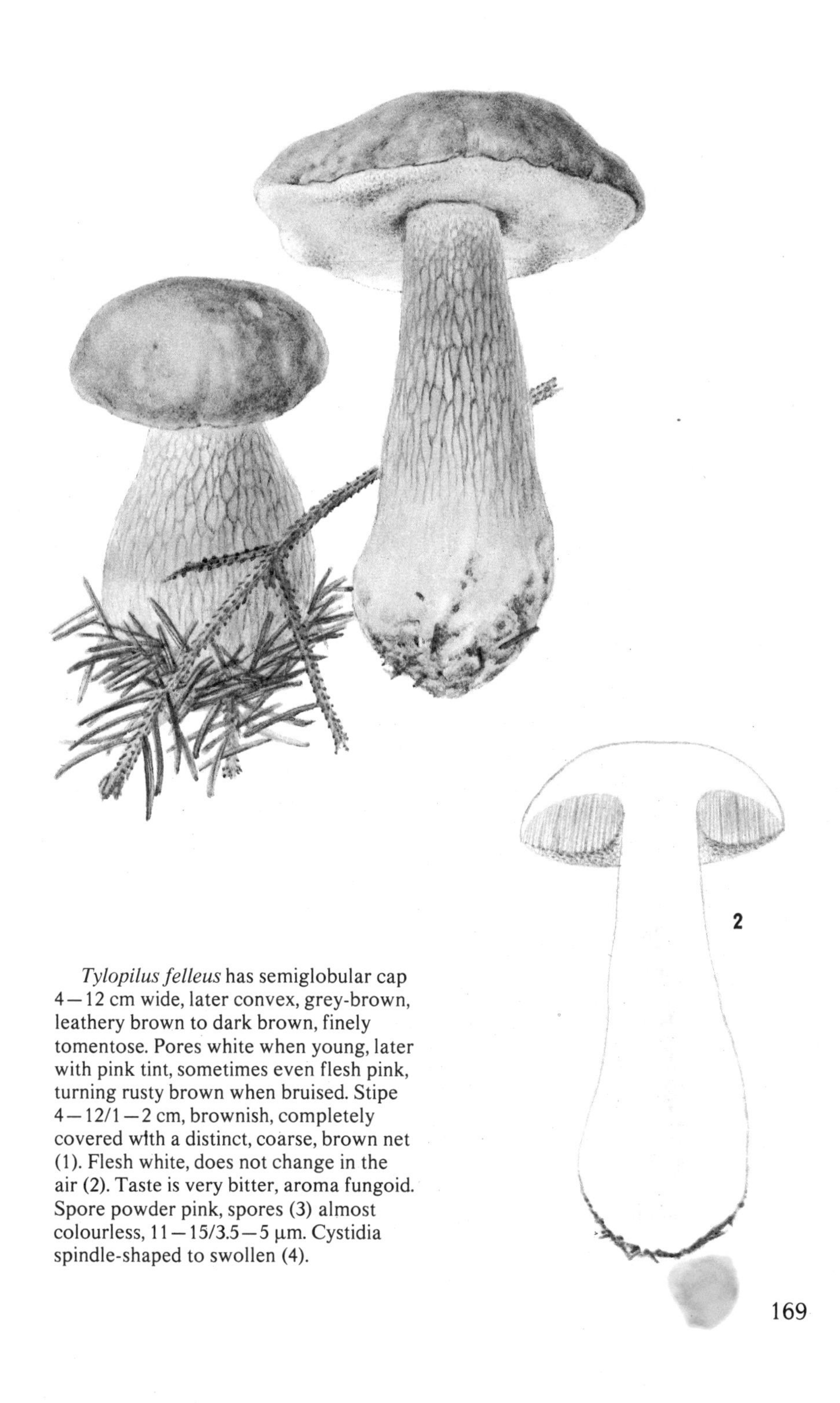


*Tylopilus felleus* has semiglobular cap 4—12 cm wide, later convex, grey-brown, leathery brown to dark brown, finely tomentose. Pores white when young, later with pink tint, sometimes even flesh pink, turning rusty brown when bruised. Stipe 4—12/1—2 cm, brownish, completely covered with a distinct, coarse, brown net (1). Flesh white, does not change in the air (2). Taste is very bitter, aroma fungoid. Spore powder pink, spores (3) almost colourless, 11—15/3.5—5 μm. Cystidia spindle-shaped to swollen (4).

## Old Man of the Woods, Cone-like Boletus *Boletaceae*

*Strobilomyces floccopus* (VAHL ex FR.) KARST.

Syn.: *Strobilomyces strobilaceus*

The Cone-like Boletus attracts attention with its brown-black fruit-bodies, coarsely scaly cap and flesh which turns black when cut. In the United States of America it is called the Old Man of the Woods, since its fruit-bodies resemble old, dying mushrooms. The Cone-like Boletus is exceptional among boleti not only in its appearance, but in the microscopic structure of its spores, which are brown and globular, with a slightly swollen wall and a distinctly patterned surface. Spores of most other boleti are yellow-brown, spindle-shaped, thin-walled and smooth on the surface. In Europe the genus *Strobilomyces* has only one representative; a closely related North American species, *S. confusus,* is smaller, with wrinkled rather than patterned spores. The genus has many tropical species. The Cone-like Boletus grows in summer and autumn in groups in coniferous and deciduous forests. It is most abundant in hills and in soils with a high acid content. In lowlands it forms mycorrhiza with beech trees; in uplands it grows under spruce and fir. It is widespread throughout the temperate zone of the northern hemisphere and has probably been introduced to South America. It is not poisonous, but its tough old stipes are difficult to digest. In Germany it is considered inedible: elsewhere in Europe it is regarded as edible but inferior. In the United States of America it is regarded as a good edible mushroom.

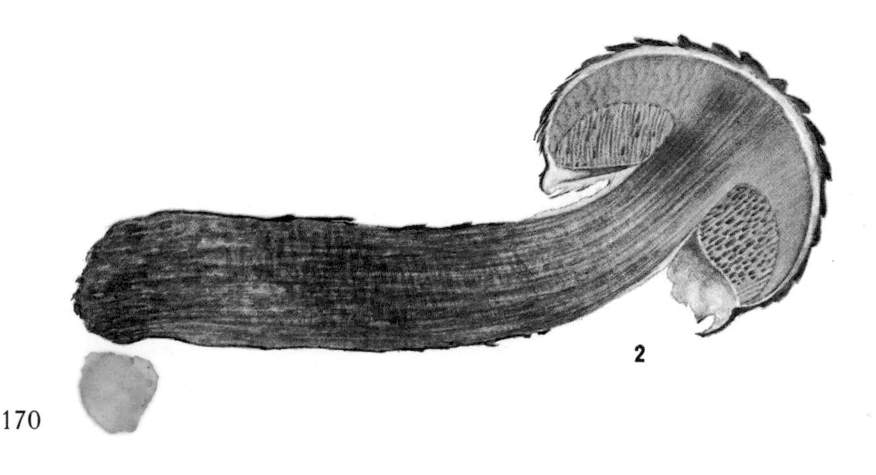

*Strobilomyces floccopus* has cap 5—12 cm wide, umber-brown or black-brown, entirely covered with large black overlapping scales. Pores whitish, covered with grey-white veil (1) when young, later grey; turns black when bruised. Stipe 7—15/1—3 cm, same colouring as cap, with coarse, filamentous scales. Flesh whitish, turns red when cut, finally becoming black-purple (2); a drop of ferrous sulphate colours it dark blue-green. Taste and smell fungoid. Spore powder black-brown; spores (3) brown, 10—13/9—10 µm.

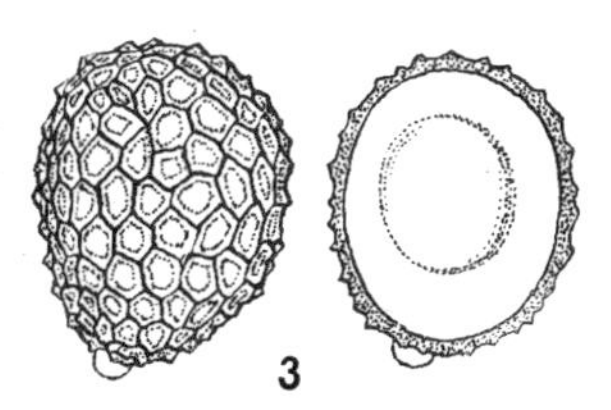

## **Orange-red Lactarius**

*Russulaceae*

*Lactarius volemus* (FR.) FR.

This representative of the genus *Lactarius,* or Milk-caps, can be cooked in various ways, but is best quickly fried in butter with caraway seeds; it can also be eaten raw with salt or in salads. *Lactarius volemus* is found in summer and autumn in coniferous and deciduous forests; it prefers lime soil and grows individually or in very small groups. It seems that it is becoming rarer, although the reasons for its gradual disappearance are not known. Though generally distributed in lowlands and highlands of the temperate zone of the northern hemisphere, it is only abundant in certain regions.

Although *Lactarius volemus* can be recognized by its red-brown or orange-yellow fruit-bodies, creamy gills and sweet, milky juice, it is often confused with other related species, for instance, the Mild Mushroom or Oak Milk-cap *(L. quietus).* This species has a brown, concentrically zoned cap and a stipe with a rusty brown base; its 'milk' and flesh taste slightly acrid. The Mild Mushroom lives in mycorrhizal association with various oak species; it likes natural oak stands or oak and elm forests and also parks and gardens. It is edible, but its flavour cannot be compared with *L. volemus.* A similar species, *L. subdulcis,* grows in beech forests. It has a brown or chestnut brown cap without zones; the stipe has the same colouring and its watery milk is also acrid. It is edible, but inferior in quality.

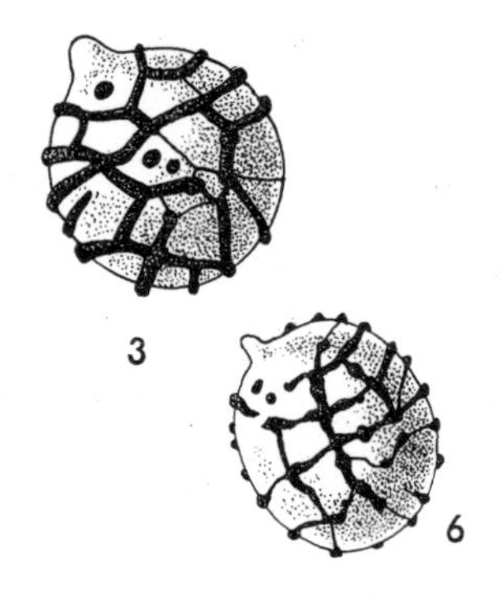


*Lactarius volemus* (1) has cap 5—15 cm wide, red-brown, orange-yellowish. Gills whitish to creamy, turning rusty. Stipe 6—12/1—3 cm, same colouring as cap. Flesh whitish; a drop of ferrous sulphate colours it grey-green. Flavour pleasant; old mushrooms smell of pickled herrings. Milk white and sweet, gradually turning brown (2). Spore powder whitish; spores (3) colourless, 7.5—10 μm.

*Lactarius quietus* (4) has cap 6—8 cm wide, dirty brown, red-brown, slightly zoned. Gills pale ochre to pale brown. Stipe 3—6/0.5—1.2 cm; same colouring as cap, rusty brown at the base. Flesh light cinnamon (5); a drop of ferrous sulphate colours it ochre yellow. Milk is yellowish-creamy, unchangeable, acrid. Spore powder creamy yellow; spores (6) yellowish, 7.5—9/6.5—7.5 µm.

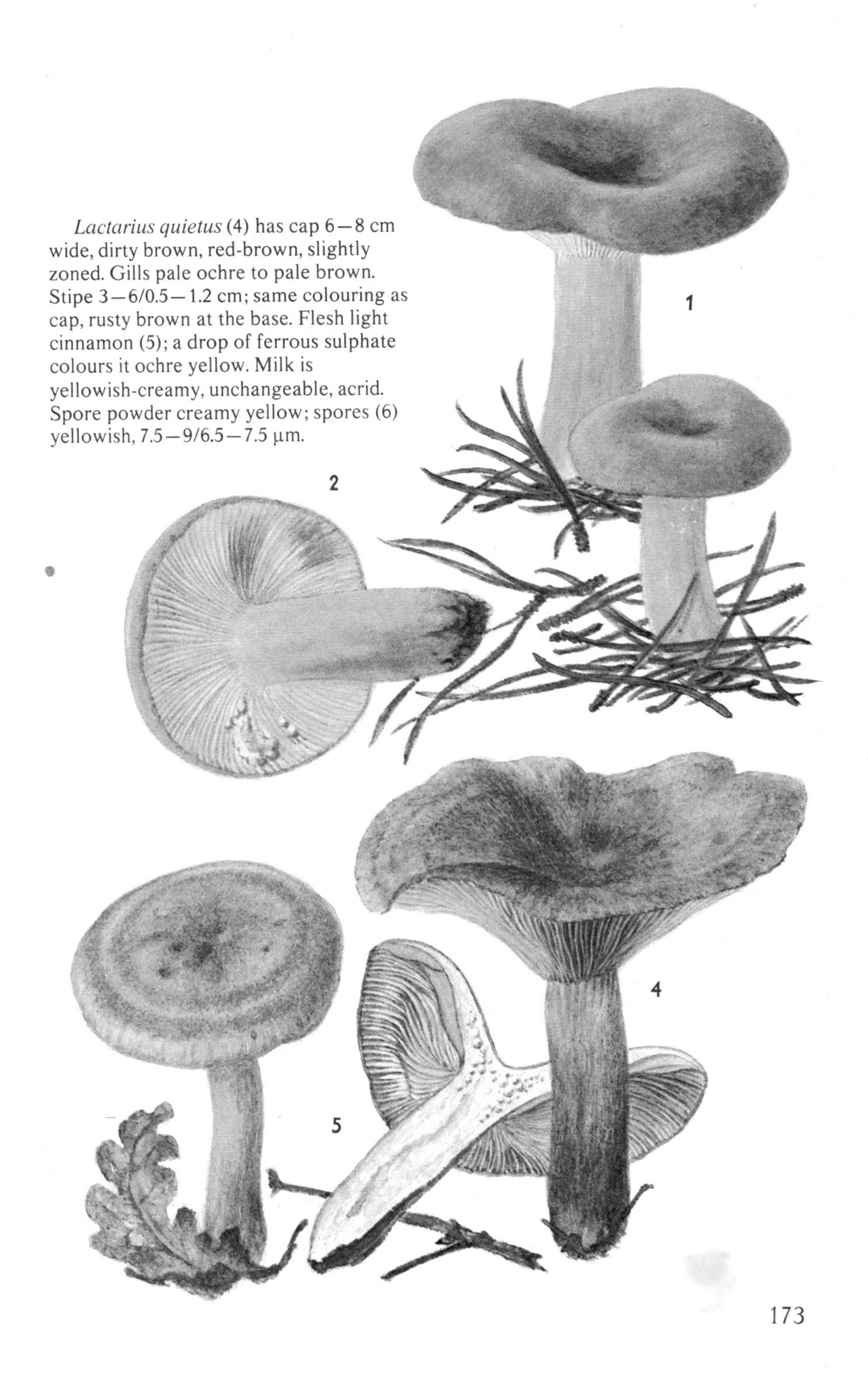

Several species of the Milk-cap genus *(Lactarius)* have such a distinctive aroma that it is quite easy to identify them by it.

Dry fruit-bodies of *L. helvus* smell of chicory or fenugreek. This Milk-cap can be found in autumn on acid soil in coniferous forests, particularly in spruce stands close to mountainous peat-bogs; it even grows in peat. It reaches as high as the subalpine zone of mountains, where it grows under dwarf pines. This species is widespread in the temperate zone of the northern hemisphere. It is lightly poisonous and therefore can be eaten only as a spicy additive in soups. A similar but smaller species, which has a strong camphor aroma, is *L. camphoratus.* It has a red-brown cap and is glabrous; its gills have the same colouring and the stipe is brown-purple. *L. serifluus,* another species with a similar aroma, has a chestnut brown cap and stipe and yellow or cinnamon yellow gills. It yields colourless milk which curdles when exposed to the air. It grows only under oak.

Although the coconut-scented Milk-cap *(L. glyciosmus)* has a pleasant and distinctive coconut aroma its flavour is too acrid to be edible. It grows in summer and autumn in large numbers in mixed woods, especially under birch, and rarely under fir trees. *L. fuscus,* which also has a coconut aroma, grows on lime soil in coniferous forests, particularly spruce. Its fruit-bodies are dark grey-brown to black-brown. It is also edible.

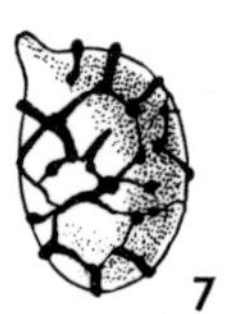


*Lactarius helvus* (1) has cap 4—16 cm wide, ochre brown, ochre red, with flake-like scales. Gills yellowish-ochre (2). Stipe is 4—13/0.5—4.5 cm, same colouring as cap. Flesh grey-ochre (3); milk clear, watery, bitterish. Spore powder cream-yellowish; spores (4) yellowish 6.5—9/5.5—6.5 µm.

*Lactarius glyciosmus* (5) has cap 2—6 cm wide, light grey, beige-ochre, finely tomentose. Gills are light ochre (6). Stipe 1.5—6/0.2—0.8 cm, pale ochre. Flesh is whitish, later beige; milk white, mild, later slightly acrid. Spore powder cream white; spores (7) colourless, 7—8.5/5.5—6.6 µm.

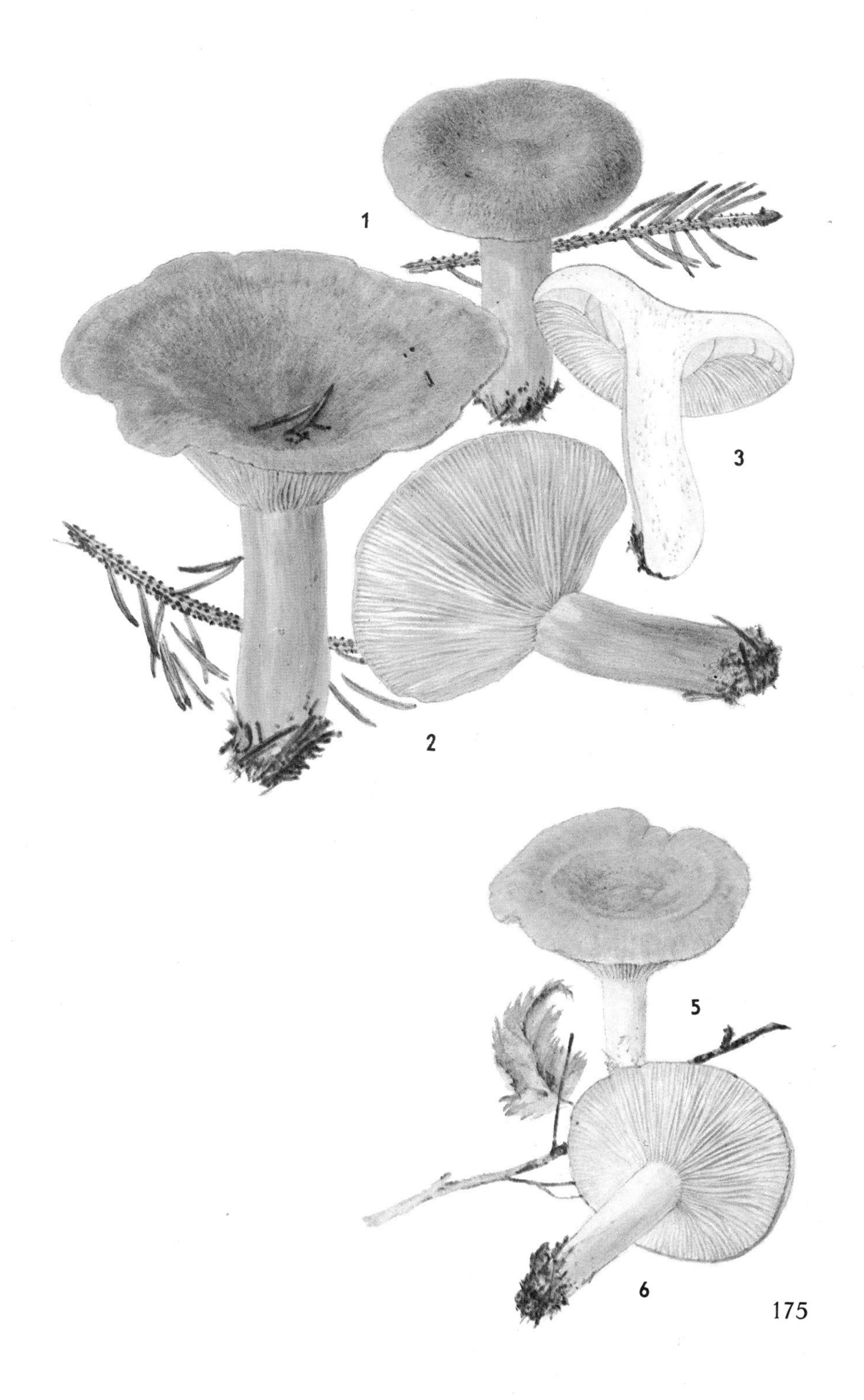
1
2
3
5
6

## Rufous Milk-cap, Red Lactarius *Russulaceae*

*Lactarius rufus* (SCOP. ex FR.) FR.

The Rufous Milk-cap is the most pungent of all 80 or so species of the Milk-cap genus *(Lactarius).* At first its taste is mild, but it soon begins to sting the mouth and continues to do so for a long time. However, it can be eaten if the muhsrooms are steeped for about 10 hours and then boiled in salt water for 10 minutes and strained before cooking. The Rufous Milk-cap can be found from the end of May to late autumn. It is very abundant, even in dry periods, on acid soil in coniferous and mixed forests; it forms mycorrhizal associations with spruce, pine and fir. It occurs in lowlands and high up in mountains as far as the subalpine zone, where it grows under dwarf pine. It occurs throughout Europe, North America and even above the arctic circle. The Rufous Milk-cap can be easily recognized by its rusty red fruit-bodies, umbo in the centre of the cap and its pungent taste. A similar species, the camphor-scented Milk-cap *(L. camphoratus),* has a brown-red cap with umbo, brownish gills, colourless watery milk and a distinct aroma of camphor; it grows in deciduous and coniferous forests. Another similar species, the Oak Milk-cap *(L. quietus)* has a red-brown cap without umbo, and a rusty brown stipe base; it grows only under oak.

*L. mitissimus* is an edible milk-cap with a slightly acrid taste when raw. It grows from summer to late autumn in coniferous forests. A closely related species, *L. porninsis,* also has a bright orange colour, fruity aroma and an acrid taste. It also grows in coniferous forests, but only under larch.

*Lactarius rufus* (1) has cap 3—8 cm wide, with a central umbo, red-brown with a silvery sheen. Gills fleshy red-ochre; stipe 4—9/0.5—2 cm, same colouring as cap. Flesh whitish, later with a fleshy tint (2). Milk is white, unchangeable, strongly acrid. Spore powder whitish.

*Lactarius mitissimus* (3) has cap 3—7 cm wide, with an umbo in the centre, orange, orange-red. Gills creamy, pale ochre; stipe 3—6/0.4—0.8 cm, orange. Flesh yellowish white (4); milk is white, unchangeable, sweet, later acrid. Spore powder cream-yellow; spores (5) of both species almost colourless, 8—9.5/6.5—7.5 μm.

## **Saffron Milk-cap**

*Russulaceae*

*Lactarius deliciosus* FR.

Most mushroom pickers think of the Saffron Milk-cap *(Lactarius deliciosus)* as an orange mushroom with green patches, which yields a carrot-red milk. In fact the name refers to several species which differ in colour and size, in the colour change of their milk and flesh in the air and in the size of their spores, and grow under different species of trees. The true Saffron Milk-cap *(L. deterrimus* (syn. *L. deliciosu* s.l.)) forms myccorhizal associations only with spruce trees. It grows in summer and autumn in large numbers in spruce stands and grassy spruce nurseries. The fresh mushrooms have a slightly bitter acrid taste and an indistinct aroma. They are inedible when raw, but are delicious when cooked.

The rarer *L. deliciosus* (syn. *L. pinicola)* lives in mycorrhiza only with pine. It can be found in summer and autumn, growing individually in pine forests in sandy or lime soil and also under solitary pine trees in parks and gardens. Other milk-caps which yield orange or wine-red milk are *L. sanquifluus, L. salmonicolor* and *L. semisanquifluus.* They differ only slightly from the illustrated species. All 'true' milk-caps are edible and have an equally good flavour, similar to that of *L. deterrimus.* However, they can be confused with the Woolly Milk-cap *(L. torminosus)* which has a woolly cap with woolly fibrils along the margin and yields an intensely acrid white milk.

*Lactarius deterrimus* (1) has cap 3—10 cm wide, light orange-ochre or almost creamy in places, without zones and with very few green patches, slimy. Gills orange-ochre (2); stipe 3—5/1—2 cm, same colouring as cap. Flesh white, milk carrot-red and unchangeable, with acrid taste (3).

*Lactarius deliciosus* (4) has cap 3—8 cm wide, fleshy or orange-red, with dark zones, slimy, with green patches when old. Gills light flesh red-yellow, orange-ochre with greenish patches (5). Stipe 3—8/1—3 cm, same colouring as cap. Flesh white, slowly becoming red; milk carrot red, very slowly becomes pale orange (6), taste acrid. Spore powder of both species light ochre; spores (7) almost colourless, 7.5—10/6—7.5 µm in *L. deterrimus* and 8.5—9/6.5—7 µm in *L. deliciosus.*

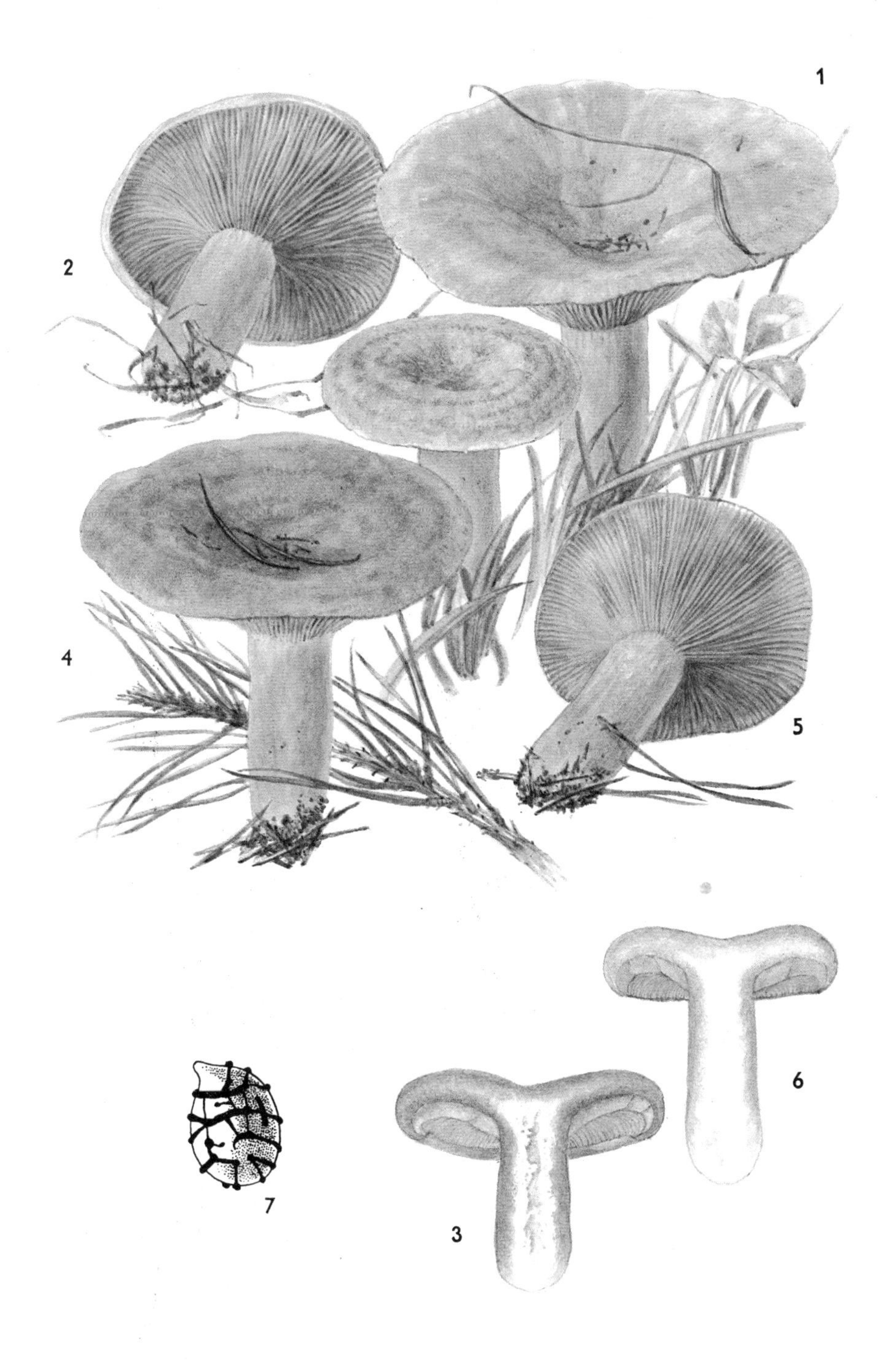
1
2
4
5
6
7
3

## Sombre Lactarid, Base Toadstool, Ugly Toadstool

*Russulaceae*

*Lactarius necator* (BULL. em. PERS. ex FR.) KARST.

Syn.: *Lactarius turpis*

Both *Lactarius necator* and *L. blennius* have a very strong acrid taste and are inedible when cooked in the usual way. They do not contain true toxins, but strong resinous irritants which affect the walls of the digestive tract. These bitter substances can be removed by boiling. In some countries in north-eastern Europe these two species are boiled in water for 10 minutes, then strained and boiled again. After this they are ready to be cooked in the usual way. But by this time they are rather inferior in taste and texture.

The Sombre Lactarid is especially abundant in coniferous forests on acid soil under spruce or pine trees, and also grows in mixed woods, where it forms mycorrhiza with birches. It is distributed throughout Europe, Asia and North America. The related *L. blennius* lives in mycorrhiza with beech species. It can be found in old parks and gardens and is characteristic of mountainous beech forests. *L. blennius* can be distinguished from *L. necator* by the colour of the cap, which is predominantly green, while in the latter it is brown. *L. blennius* has a smooth cap margin, while in *L. necator* it is tomentose. The colour of the gills also differs; in *L. necator* it is yellowish grey-green from young, while in *L. blennius* they remain white for a long time, becoming grey-green only when old.

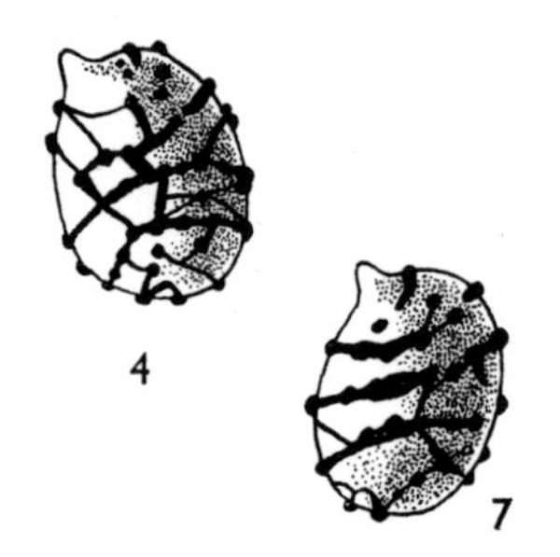


*Lactarius necator* (1) has cap 5—20 cm wide, dark olive green to brown, viscous, often with dark patches, and an enrolled, woolly margin with a lighter shade. Gills dirty yellowish even when young (2). Stipe 3—8/1—1.5 cm, greenish, dirty olive; flesh white, later brownish (3). Milk white, unchangeable, acrid.

*Lactarius blennius* (5) has cap 3—8 cm wide, green, grey-green, with brown patches, slimy, smooth. Gills white when young, yellowish grey-green when old (6). Stipe 3—6/0.8—1.5 cm, same colouring as cap. Flesh whitish to greyish; milk white, grey-green when dry, acrid. Spore powder of both species cream yellow; spores colourless; in *L. blennius* (7) 7.5—8/6 μm, in *L. necator* (4) 6—8/5.5—6.6 μm.

## Woolly Milk-cap, Griping Toadstool *Russulaceae*

*Lactarius torminosus* (SCHFF. ex FR.) S. F. GRAY

The Woolly Milk-cap grows from summer to autumn in and outside forests, exclusively under birch trees, throughout the temperate zone of the northern hemisphere, even as far north as Alaska. It is the only Milk-cap species which is considered poisonous. The highly acrid terpene toxids of the raw mushroom irritate the intestines and within 2 to 4 hours cause vomiting, stomachache and diarrhoea. The sickness stops in two days. But if the mushrooms are boiled in salt water for at least 15 minutes and then strained, they are harmless and can be safely cooked and eaten. In some parts of the Soviet Union the Woolly Milk-cap is preserved in salt or conserved by fermentation and then served with soured cream and vodka. In Finland it is one of the most popular mushrooms.

The Woolly Milk-cap is sometimes confused with *L. deterrimus* which, however, never has a tomentose cap or white milk. The Woolly Milk-cap is much more acrid than *L. deterrimus.*

*L. pubescens* occurs during the same season in damp habitats. It is also acrid, but has a geranium-like scent. It can be recognized by its paler, beige cap, which has no zones.

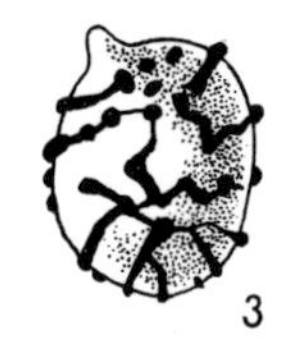


*Lactarius torminosus* (1) has cap 5—13 cm wide, fleshy pink to fleshy brown, with dark circles and inrolled, pubescent margin. Gills creamy pinkish; stipe 4—8/1—2 cm, same colouring as cap. Flesh whitish, later pinkish (2); milk white, unchangeable, acrid. Spores (3) colourless, 7.5—10/6—8 μm.

*Lactarius pubescens* (4) has cap 2.5—9 cm wide, beige or pale beige, without zones, with pubescent margin. Gills whitish or yellowish (5); stipe 2—5.5/0.3—1.2 cm, same colouring as the cap. Flesh white, unchangeable; milk white, unchangeable, acrid. Spore powder of both species creamy yellow; spores (6) colourless, 6.5—8.5/5.5—6.5 μm.

1
2
4
5
6

The white milk-caps include about five species, all of which have a very acrid taste. This unpleasant flavour can be eliminated by cooking and the mushrooms thus become harmless. In parts of the Ukraine, Rumania and Poland *L. piperatus* is picked and sold in markets; it has quite a pleasant taste when fried with bacon and onions. The acrid taste also disappears when the mushroom is fermented. *L. vellereus* and *L. piperatus* are very abundant in summer and autumn, even when the weather is dry. Both species can be found in deciduous forests, where they form mycorrhizal associations with beech and oak; less commonly they are found in coniferous forests under spruce and pine trees. They are widely distributed in the temperate zone of the northern hemisphere.

*L. vellereus* is characterized by its velvety tomentose cap, sparse gills and short but stout stipe. *L. piperatus* has a glabrous cap with densely crowded gills and a long stipe. This species resembles other less common white milk-caps, namely *L. glaucescens* and *L. pergamenus,* whose milk becomes bright yellow if a drop of potassium hydroxide is added, whereas the milk of *L. piperatus* remains unchanged. When the fruit-bodies of both species are drying, they display grey-green patches. *L. glaucescens* has shortly decurrent gills and yields white milk, which curdles and turns green in the air. The gills of *L. pergamenus* are not decurrent and its white milk does not curdle but turns slightly yellow. Another relative of *L. vellereus* is *L. deceptivus,* which grows in North America. It differs from the above-mentioned species in having a woolly margin to its cap, and large spores.

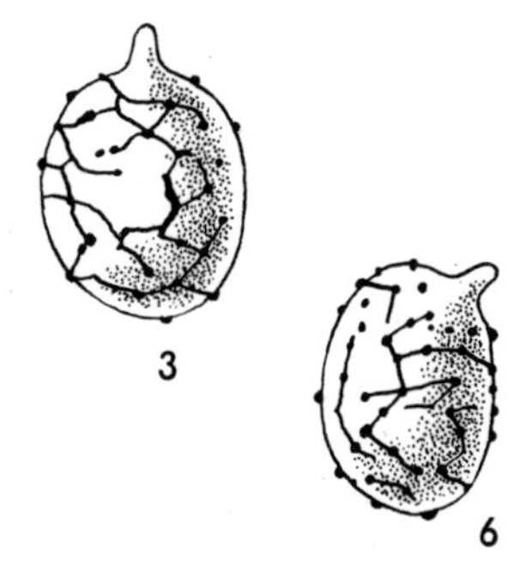


*Lactarius vellereus* (1) has cap 8—30 cm wide; stipe 2—6/2—5 cm, white and softly tomentose. Gills sparse, white, ochre when old (2). Flesh white, late pale yellowish. Milk white, acrid. Spores (3) colourless, 7.5—9/8 µm.

*Lactarius piperatus* (4) has cap 5–30 cm wide; stipe 3–9/1–3 cm, white, glabrous. Gills creamy, crowded, narrow and decurrent (5). Flesh white, later yellow; a drop of ferrous sulphate colours it beige-pink. Milk white, later pale yellow, acrid, does not curdle. Spore powder of both species whitish; spores (6) colourless, 8–9.5/5–5.7 µm.

## Stinking Russula

*Russulaceae*

*Russula foetens* FR.

Mushrooms of the genus *Russula* are distinguished by the fragile, apple-like texture of the stipe, which is not filamentous as in most mushrooms. The end of a pen-knife will dig out a rounded piece rather than separating the stipe into filaments. The texture of the stipe results from the presence of globose cells (spherocystidia) located between long hyphae.

The Stinking Russula is characterized by its light brown, slimy cap, which is grooved along the margin, and above all by its repulsive smell and extremely acrid burning taste. It grows in all forests in summer and autumn, even in the mountains. A similar species, *R. laurocerasi,* is smaller and has a pleasant, bitter almond aroma; however it is also acrid and therefore inedible. It is abundant on lime soil from June to October in deciduous forests, under oak and beech species.

The Yellow Russula *(R. ochroleuca)* is one of the most common russulas growing in European forests. In summer and autumn it forms large colonies in coniferous forests of submountainous and mountainous regions and is characteristic of natural mountain spruce stands. It can also be found in lowland oak and birch woods. It prefers acid soil and is intolerant of lime. The Yellow Russula is only slightly acrid, but is nevertheless inedible. The similar *R. fellea* has an ochre-yellow rather than white stipe. It has a pleasant honey aroma, but a very bitter taste, and grows mainly in beech forests.

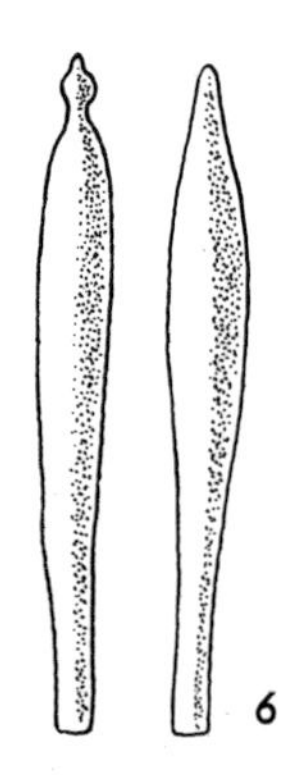


*Russula foetens* (1) has cap 5—15 cm wide, ochre, light brown, slimy, with lobed margin. Gills straw yellow, turning rusty in patches when bruised and old. Stipe 5.5—11/1.2—5 cm, white, later turning rusty. Flesh pale yellowish (2). Spore powder pale creamy; spores (3) colourless, 8—10/6.5—8.5 µm.

*Russula ochroleuca* (4) has cap 4—10 cm wide, yellow, yellow-ochre, sticky. Gills white, later pale creamy. Stipe 3.5—7/1—2.5 cm, white, slightly yellowish. Flesh is white; a drop of ferrous sulphate colours it pink. Spore powder whitish; spores (5) colourless, 8—10.5/6.5—8.5 µm. Cystidia of both species are spindle-shaped (6).

*R. vesca* and *R. mustelina* are among the tastiest of mushrooms. When raw, they have a sweet, pleasantly nutty flavour; they can be cooked in many ways. The two species differ in their cap colour, which is red-brown in *R. vesca* and brown in *R. mustelina,* and in their gill colour, which is white in *R. vesca* and creamy in *R. mustelina. R. vesca* grows in summer and autumn in dry oak, pine and beech forests and along their edges. It is absent in peat-bogs and mountains. *R. mustelina* prefers mountainous coniferous forests, where it thrives in summer and autumn under spruce and fir. Young specimens of *R. mustelina* resemble the Edible Boletus in colour and overall appearance. Both species are distributed throughout Europe and Asia; *R. vesca* also grows in North America.

The Fork-gilled Russula *(R. heterophylla),* distinguished by its green cap, occurs in the same habitat as *R. vesca.* Mountainous coniferous forests are the home of another relative, *R. integra,* which grows under spruce and fir. It has a red-brown cap and its mature gills are ochre yellow. Though it is edible, its flavour is inferior to *R. vesca* and *R. mustelina.*

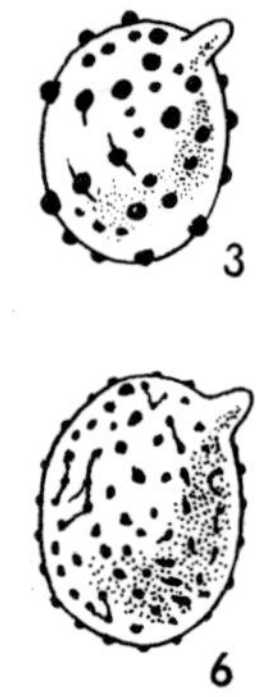


*Russula vesca* (1) has cap 6—10 cm wide, flesh red, brownish. Gills are white, rusty when old (2). Stipe 5—10.5/1.3—2.5 cm, white, rusty at the base. Flesh white; a drop of ferrous sulphate colours it orange. Spore powder white; spores (3) colourless, 6—8.5/5—6.5 µm.

*Russula mustelina* (4) has cap 5—13 cm wide, ochre, ochre-brown. Gills whitish, later creamy (5). Stipe 3—10/1.5—3 cm, white, sometimes with rusty patches. Flesh white, gradually becoming light brown; a drop of ferrous sulphate colours it red-orange. Spore powder creamy; spores (6) colourless, 6—8/5—7 µm.

2
1

4
5

# Blue Russula



*Russula cyanoxantha* SCHFF. ex FR.

It is very useful to be able to identify the various *Russula* which abound in European forests, as many of them have an excellent taste. The drawback is that many of them are very similar in appearance. One of the best-flavoured species, *R. cyanoxantha,* has a widely variable coloration; the cap can be brown-red, dark green, purple or blue. Sometimes one colour prevails (green in the variety *peltereaui*) or several colours are merged together (blue and purple in the variety *cyanoxantha).* The only reliable feature for identification is the unusually supple gills, which are greasy to thc touch. All the other *Russula* species have fragile gills which break easily. *R. cyanoxantha* grows from June to October and it is abundant in mixed and deciduous forests under oak and beech in lime-rich or acid soil, and is distributed through the whole temperate zone of the northern hemisphcre. It can be confused with several similar edible *Russula* species, but none of them has supple, unbreakable gills. And while the flesh of the other species turns orange or pink with a drop of ferrous sulphate the flesh of *R. cyanoxantha* does not change its colour. Similar species include *R. heterophylla,* which has a green shaded cap, and grows mainly in beech forests; there is also *R. olivacea,* with an olive-green or purple cap, shrivelled along the margin and with yellowish gills, and *R. vesca,* with a red-brown or cloudy red cap.

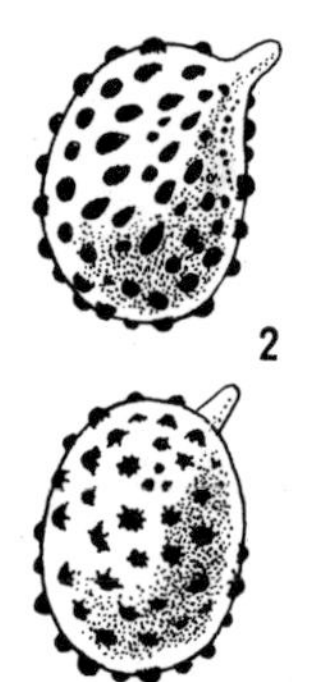


*Russula cyanoxantha* has cap 4—14 cm wide, sticky and glossy with a mixture of green, purple grey-purple and blue shades. Gills white, supple, unbreakable (1); stipe 4—10/1.5—3 cm, sometimes with a faint blue tinge. Flesh white; flavour pleasant, aroma indistinct. Spore powder white; spores (2) colourless, 6.5—10/5.5—6.5 µm. Cystidia (3) almost spindle-shaped or cylindrical.

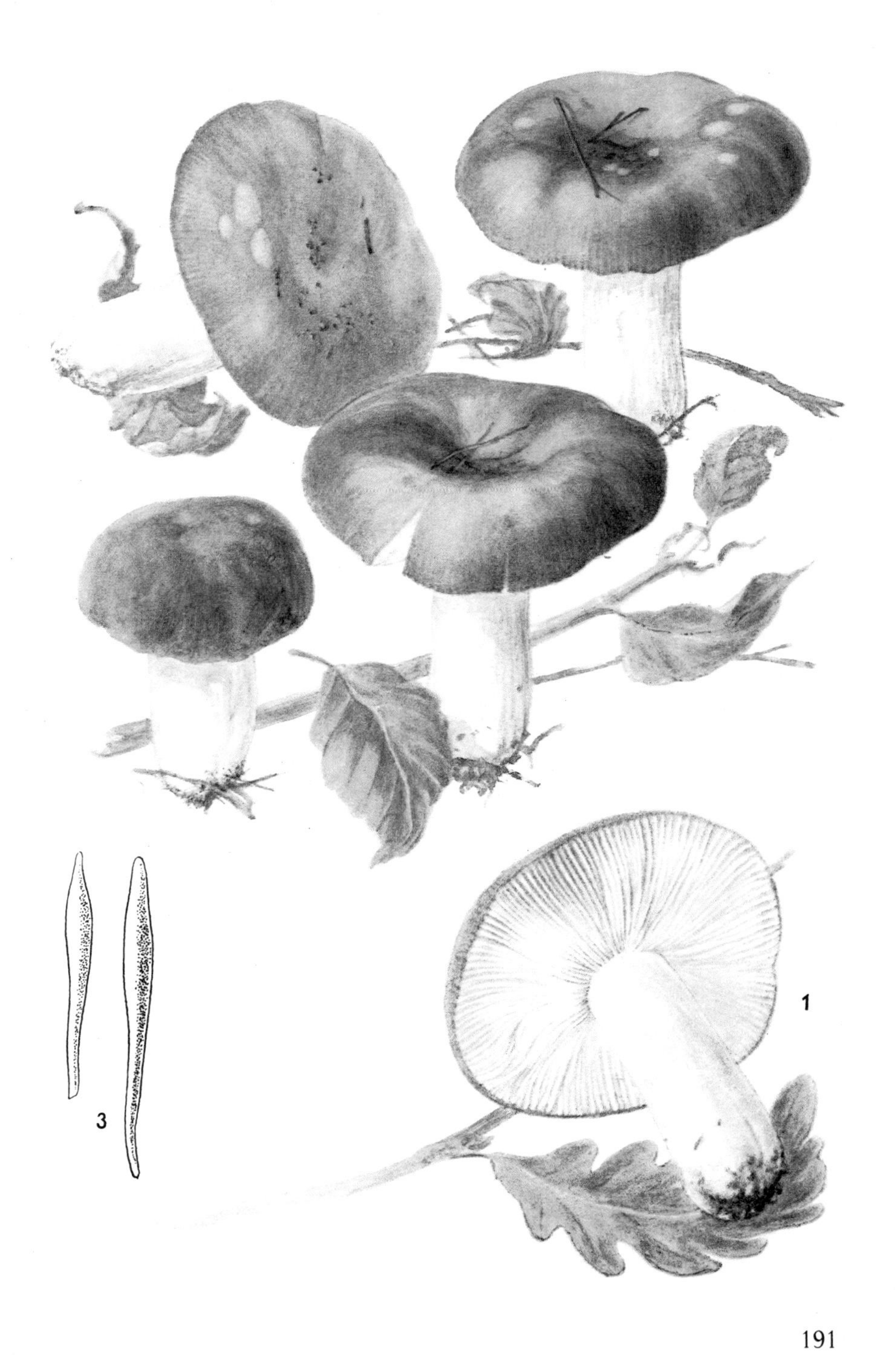
1
3

## Green Russula

*Russulaceae*

*Russula virescens* (SCHFF. ex ZANT.) FR.

The Green Russula is the best edible representative of its genus, comparable in flavour with the boleti, and can be cooked in many ways. It grows from June to October in all types of forest, particularly under oak, beech and birch in dry, light deciduous woods. It can be found in lowlands and uplands, but rarely high in mountains, and is widespread throughout Europe, Asia and North America. Its copper green cap is tomentose when young and regularly cracked into areolate patches later; this serves as an identifying feature which differentiates it from other green russulas.

The related *R. aeruginea* has a smooth, wet-looking, glossy, grass-green cap and, though edible, a slightly acrid taste. It abounds in deciduous forests under birch trees, and is found, though rarely, in coniferous forests. Its distribution is similar to that of the Green Russula, but it is rare in southern Europe. Unfortunately these two *Russula* species can be confused with the deadly poisonous Death Cap. However, there are many differences between these species; the most important is the absence in these Russula of the ring and of the swelling at the base of the stipe, which furthermore is not enveloped by a volva. Confusion with the Death Cap can therefore be easily avoided by examining the complete mushroom before cutting away from the base of the stipe.

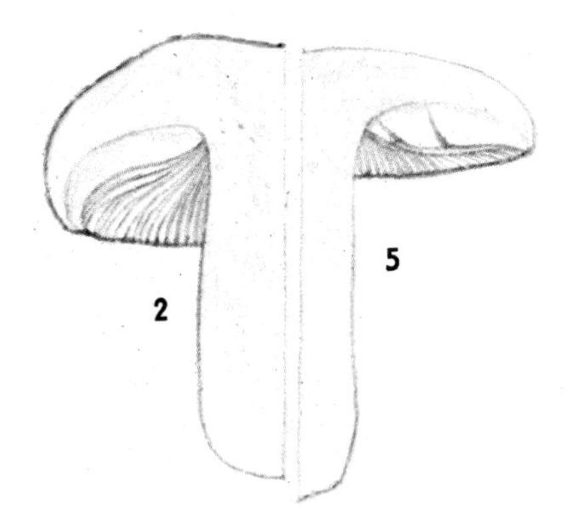


*Russula virescens* (1) has cap 5—15 cm wide, copper green, ochre-green, with a cuticle regularly cracked into reticulate patches in mature specimens. Gills white, pale creamy; stipe 2—9/1.5—4 cm, white. Flesh white (2), becoming rusty; a drop of ferrous sulphate colours it pink. Spore powder whitish; spores (3) colourless, 8—10/7—8 µm.

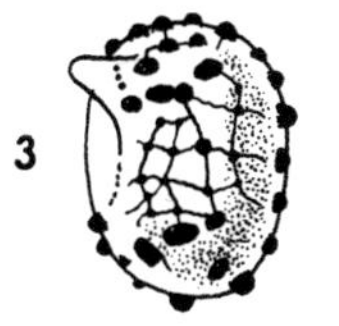

*Russula aeruginea* (4) has cap 4–10 cm wide, grass or olive green. Gills whitish, later creamy; stipe 5–8/0.8–2 cm, turning rusty at the base. Flesh white (5); a drop of ferrous sulphate colours it pink. Spore powder creamy; spores (6) colourless, 6–10/5–7 µm.

## Black Russula

*Russulaceae*

*Russula nigricans* (BULL. ex MÉRAT) FR.

The large, tough fruit-bodies of the Black Russula can be found in summer and autumn in all types of forest, but mainly in deciduous woods. It is edible but of inferior quality, and therefore is best mixed with other mushrooms. The Black Russula, *R. albonigra, R. adusta* and *R. densifolia* form a natural group of russulas called Nigricantinae, characterized by flesh which turns black when cut and by their firm and hard fruit-bodies, which in maturity are whitish to brownish. The dead mushrooms decompose very quickly and turn black, as if carbonized, lasting till the following year. The Black Russula is the largest representative of this group; its cap measures up to 20 cm in diameter and its gills are almost 2 mm thick. Its flesh when cut turns at first red and then black. *R. albonigra* has a whitish cap when young, but when bruised or old it turns black; the flesh turns black immediately when cut. It is quite rare; it grows in autumn in deciduous forests under birch trees. *R. densifolia* has a brown or black cap and very crowded gills which turn red and subsequently black when bruised, as does the flesh. It grows in summer and autumn in lime-rich soil in coniferous forests. It is not edible because of its acrid taste. The last member of the Nigricantinae group is the edible *R. adusta.* It has a greyish to grey-brown cap and its flesh turns grey when cut; it subsequently turns brown, but never completely black. It lives in all types of forest, but it prefers conifers.

*Russula nigricans* has cap usually 8—15 cm wide, whitish, later greyish, white-brown, to brown-black and black when old. Gills white, yellowish, later light grey-brown, very thick and sparse. Stipe 2—8.5/1—4 cm, white, later grey-brown. Flesh white, later brick-red (1); it turns darker (2) and finally becomes completely black; a drop of ferrous sulphate colours it pink at first and green later. Taste mild, aroma fruity or earthy. Spore powder white; spores (3) colourless, 6—8/6—7 µm.

1
2

## The Sickener, Emetic Russula



*Russula emetica* (SCHFF. ex FR.) S. F. GRAY

The Sickener is inedible because of its extremely acrid taste; it may be that the mushroom is poisonous and contains the alkaloid muscarine. The colour of its cap and its ecological requirements are strongly variable. These blood or pink-red mushrooms grow in acid soil on peat-bogs and in pine and spruce forests. The most common variety, *sylvestris*, has a vivid cherry red cap and grows in deciduous and coniferous forests. The variety *betularum* has a pink-red cap and grows under birch. Other varieties are less common and less important. The Sickener is widespread throughout the temperate zone of the northern hemisphere and with the exception of the tropical zone also in the southern hemisphere. In the colour of its cap and its strong acrid taste it resembles several other species. One is the common *R. mairei* with its carmine cap, which is pink under the cuticle; it can be found in beech woods among fallen foliage and on rotten beech stumps. Another, *R. luteotacta*, has the same cap colour as the Sickener, but its gills and stipe turn deep yellow when touched.

*R. paludosa* is characterized by its bright red or orange cap which resembles a red apple or a ripe strawberry; it has butter-yellow gills and a mild taste. It grows in mountainous coniferous forests, mainly in spruce thickets, in mountainous peat-bogs and also in pine forests. It is less common than the Sickener, but it is edible.

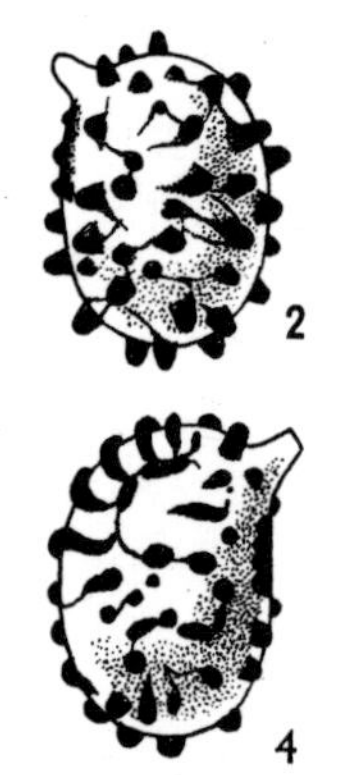


*Russula emetica* (1) has cap 5—10 cm wide, blood or pink-red, glossy. Gills white; stipe is 5—8/1.5—2 cm, white. Flesh white, unchangeable. Spore powder white; spores (2) colourless, 7.5—12.5/6—9.5 μm.

*Russula paludosa* (3) has cap 8—15 cm wide, strawberry or orange red, glossy. Gills white, later butter-yellow. Stipe 5.5—10/2—3.5 cm, white, pinkish; flesh white. Spore powder ochre; spore (4) colourless, 8—11/6.7—8.5 μm. Cystidia of both species spindle-shaped (5).

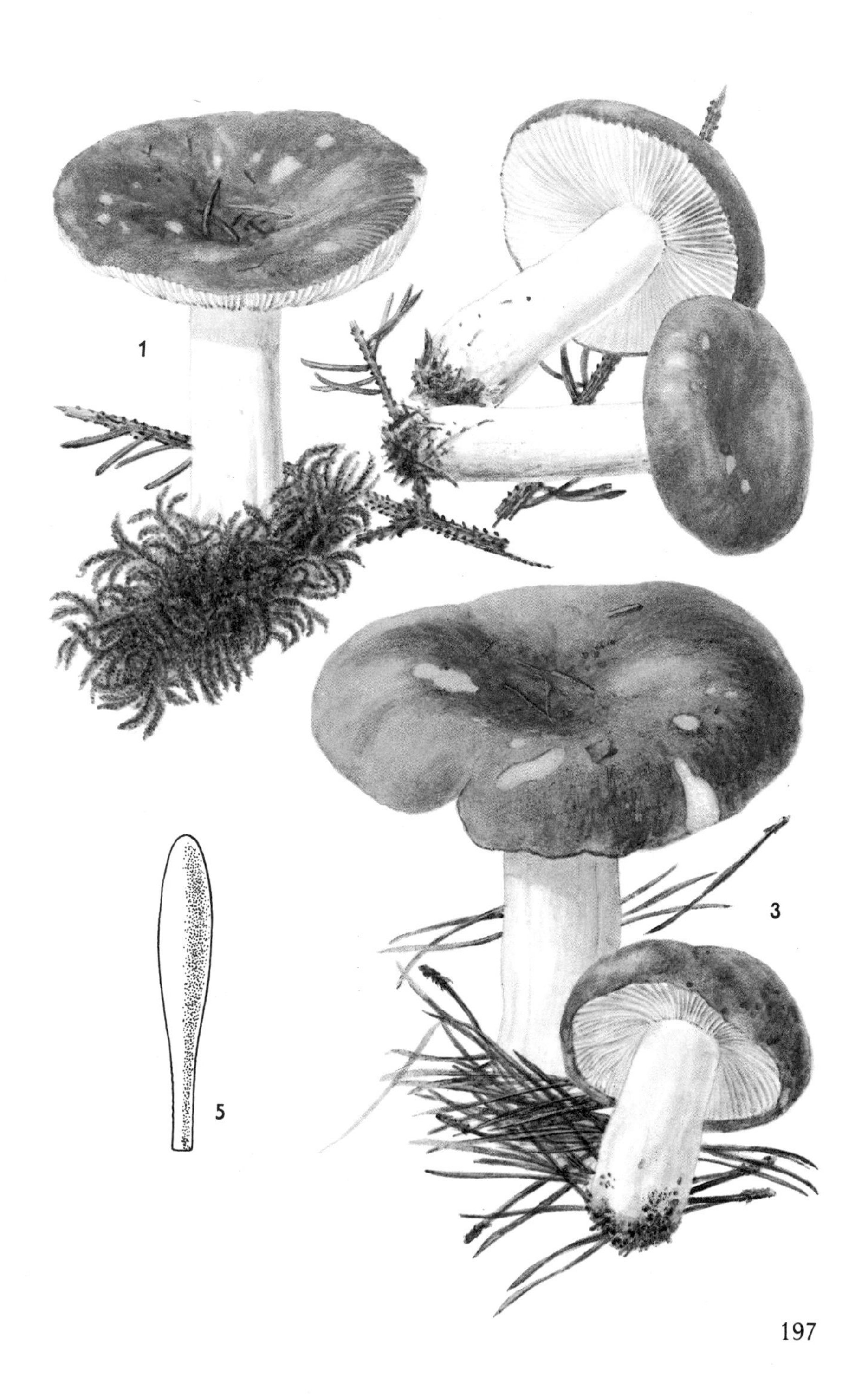
1
3
5

## **Common Stinkhorn, Wood Witch** *Phallaceae*

*Phallus impudicus* L. ex PERS.

The Stinkhorns, which belong to the *Gasteromycetes,* are generally widespread in tropical and subtropical zones. The Common Stinkhorn, on the other hand, is absent from the tropics, but abounds in the temperate zone of the northern hemisphere in the area stretching from Japan to North America; it also is found in South America and Australia. It is most abundant in Europe. Its remarkably phallic shape, and the repulsive smell of the mature mushroom, have attracted popular interest for many centuries. Because of its shape the ancient Romans dedicated it to Ceres, goddess of agriculture and fertility. In the Middle Ages it was used in the preparation of love potions. Because of the stinking slime on its cap it was considered poisonous, but the ovoid young fruit-bodies ('eggs') are edible and have a fish-like taste when fried in fat. The spores of the Common Stinkhorn and other species of the Phallaceae family are not scattered by the wind, but by insects. The insects, attracted by the smell, settle on the cap and sip the malodorous mucus together with the spores, which stick to their bodies. The spores also penetrate the insects' digestive tracts, pass through them and are dispersed through their droppings. When the layer of mucus disappears, the fruit-bodies cease to smell. The Common Stinkhorn can be found in summer and at the beginning of autumn in humus-rich, deciduous, particularly beech forests.

The related *Ph. hadriani* is characterized by its ovoid fruit-body and volva which turns red. It grows only in Europe, northern Africa and Asia, most frequently in the sand of the seashore. The species is probably absent in North America, but it is replaced there by *Ph. ravenelii,* which is smaller and has a smooth, thimble-shaped cap.

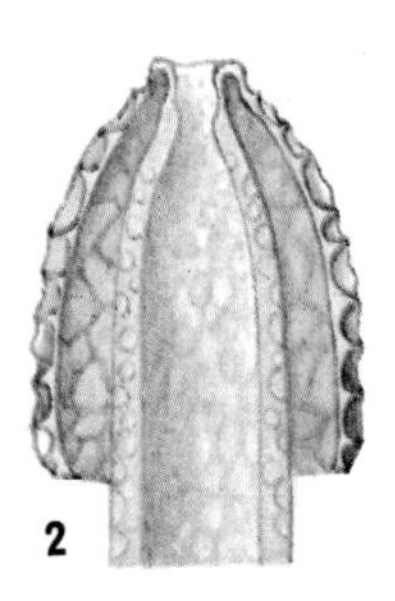
2

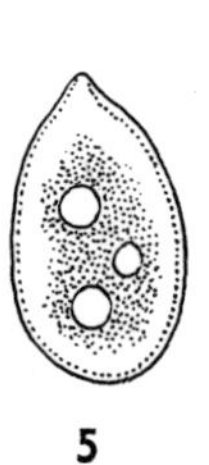
5

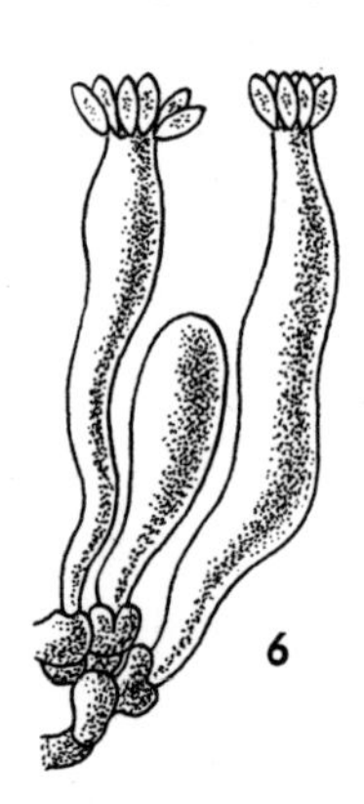
6

*Phallus impudicus* has fruit-bodies 4—6/3—5 cm in size; young individuals are globular or ovoid, whitish, with stringy mycelium at the base (1). In mature fruit-bodies the veil breaks into 2—3 lobes. The volva formed in this way carries a white, porous, hollow stipe, 10—30 cm tall and 3—5 cm wide. The ribbed cap (2) has a thimble-like shape and is covered with green, stinking slime (3) which contains spores. In old specimens the cap has no slime and is ochre to creamy (4). Taste is nondescript; smell of young specimens resembles radishes, later has a repulsive, decaying character. Spores (5) pale yellowish, 3.5—5/1.5—2 μm; 6—8 spores to a basidium (6).

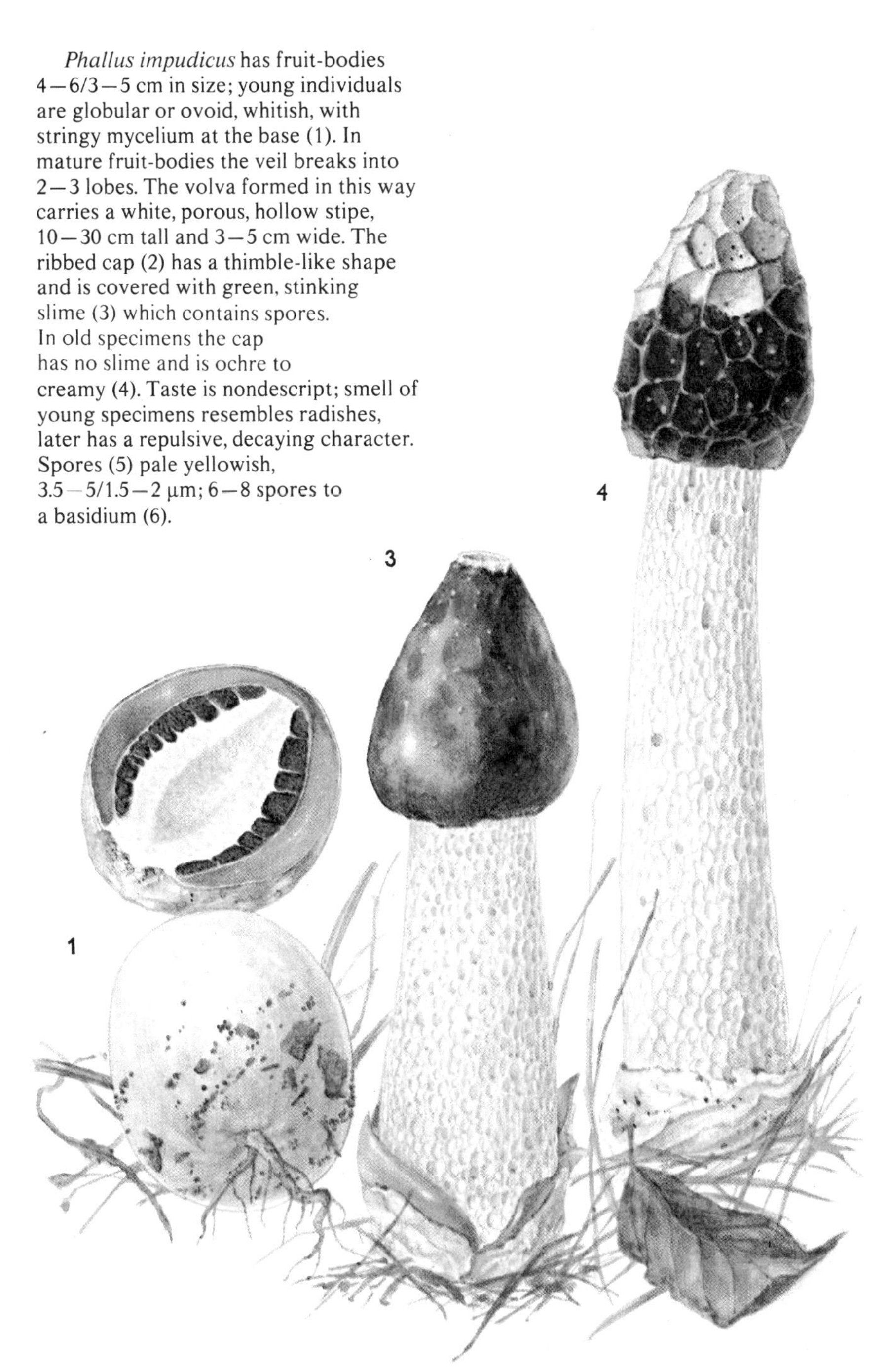

## Common Earth-ball

*Sclerodermataceae*

*Scleroderma citrinum* PERS.
Syn.: *Scleroderma vulgare*

*Scleroderma citrinum* and the related *S. verrucosum* are distributed worldwide and form mycorrhizal associations with coniferous and deciduous trees. The Common Earth-ball grows in groups in all type of woodland and is most abundant along their warm, sunny edges; it particularly likes the sandy and clay soil of pine forests. Both species form fruit-bodies in summer and autumn. *S. verrucosum* likewise grows in groups in similar habitats, but it is more abundant in places affected by human activity, along roads and tracks, in ditches and places where the original vegetation has been replaced by another type of flora. It is easy to distinguish one from another; the fruit-bodies of *S. citrinum* are covered with peridium (casing) 1—2 mm thick, which has a cracked surface and is sessile. The fruit-bodies of the Common Earth-ball sometimes host a parasitic boletus species, *Xerocomus parasiticus. S. areolatum,* a relative of *S. verrucosum,* is abundant in coniferous and deciduous forests. Its globular fruit-bodies are only 1—4 cm in size and the peridium is very thin and delicately squamous. A rare species, *S. meridionale,* can be found in the sand of seashores.

All earth-balls are slightly poisonous. Nevertheless the strong spicy flavour of the young fruit-bodies is used as spice in soups and sauces.

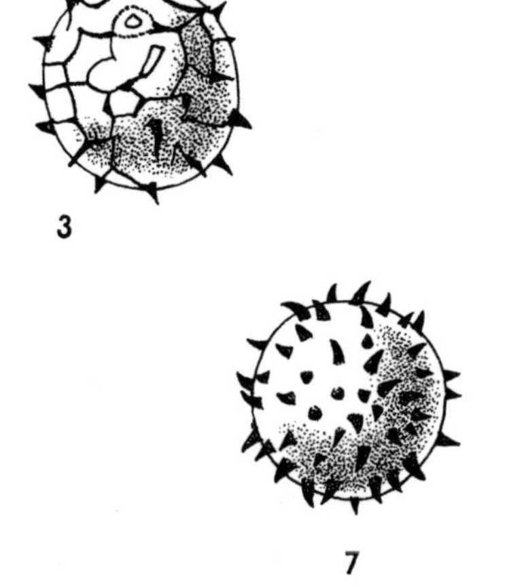


*Scleroderma citrinum* (1) has fruit-bodies 3—12 cm, irregularly globular. Peridium 2—3 mm thick, ochre-yellow or yellow-brown to brown, often with regularly cracked surface. Fruit-body (gleba) white inside at first, later pinkish, bluish to black (2). Peridium cracks in ripe specimens to make an irregular pattern. Taste mild, smell aromatic and spicy. Spore powder green-brown; spores (3) black-brown, 8—15 µm.

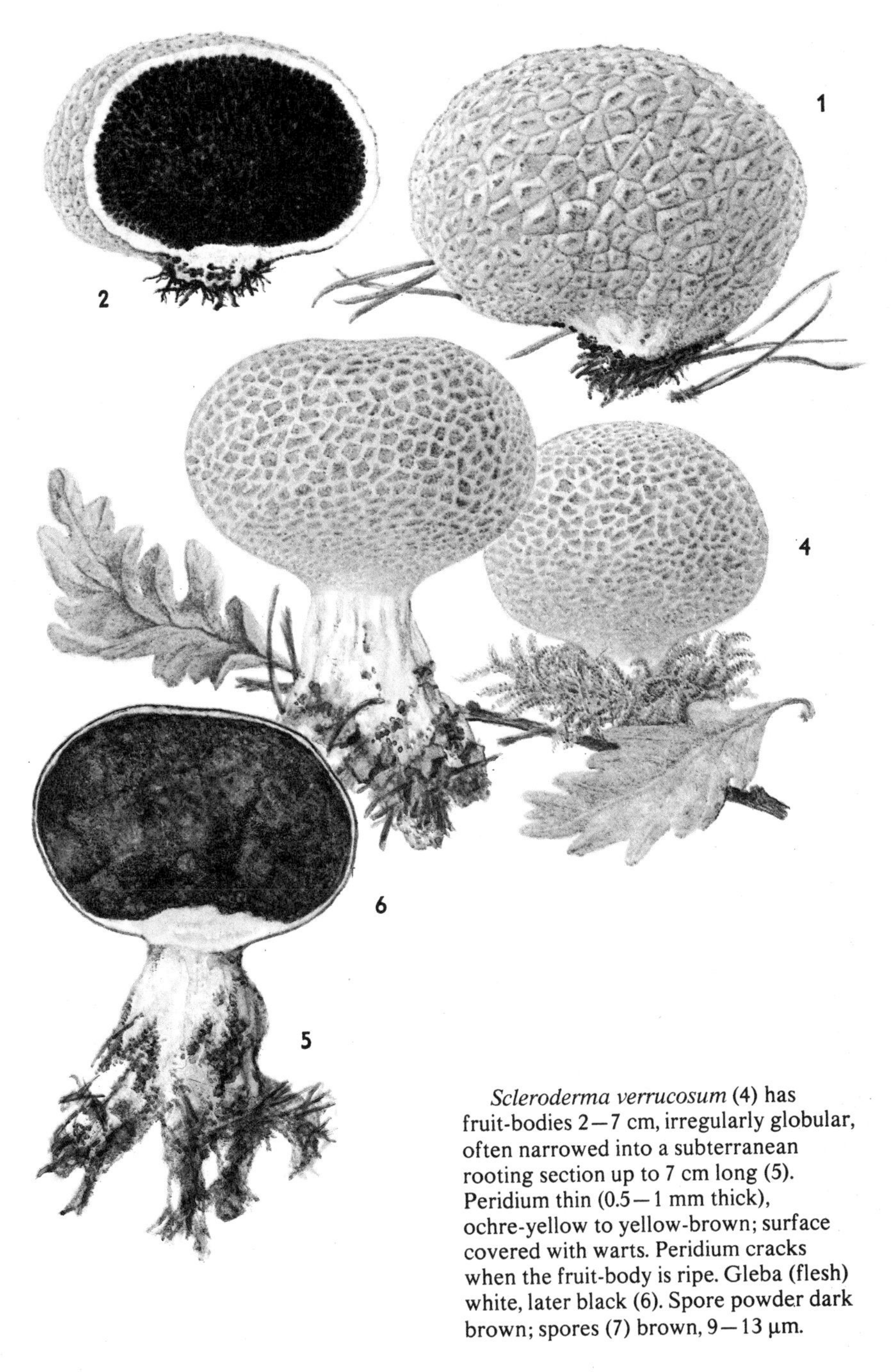


*Scleroderma verrucosum* (4) has fruit-bodies 2—7 cm, irregularly globular, often narrowed into a subterranean rooting section up to 7 cm long (5). Peridium thin (0.5—1 mm thick), ochre-yellow to yellow-brown; surface covered with warts. Peridium cracks when the fruit-body is ripe. Gleba (flesh) white, later black (6). Spore powder dark brown; spores (7) brown, 9—13 μm.

## Common Puff-ball

Lycoperdaceae

*Lycoperdon perlatum* PERS. ex PERS.

The genera *Lycoperdon* and *Calvatia* closely resemble each other; their fruit-bodies are sub-globose with an upper, fertile tissue called the gleba and a lower part which is infertile. The genera are not rich in species, in Europe the genus *Lycoperdon* is represented by 15 species and *Calvatia* by 6 species. All species are edible and have a pleasant flavour, as long as the fertile layer of the young fruit-bodies is pure white.

The Common Puff-ball is distributed worldwide. It grows in summer and autumn in small groups in coniferous and deciduous forests, and in pastures and meadows, from lowlands to highlands. The related *L. foetidum* is more common in mountains. Its exoperidium is composed of numerous spines, 2—3 mm long, which form star-shaped groups; after falling off they leave behind an areola, a circular pattern on the inner layer (endoperidium).

*Calvatia utriformis* grows only outside forests, in fertilized meadows and grazing grounds. In North American prairies and meadows it grows together with an abundant species, *C. cyanthiformis,* which has a purple exoperidium and purplish endoperidium. The latter is also a rare inhabitant of the European rocky steppes. *Lasiosphaera gigantea* also occurs on grazing ground, in gardens and deciduous forests. Its globular fruit-bodies are up to 50 cm in diameter and weigh as much as 15 kg.

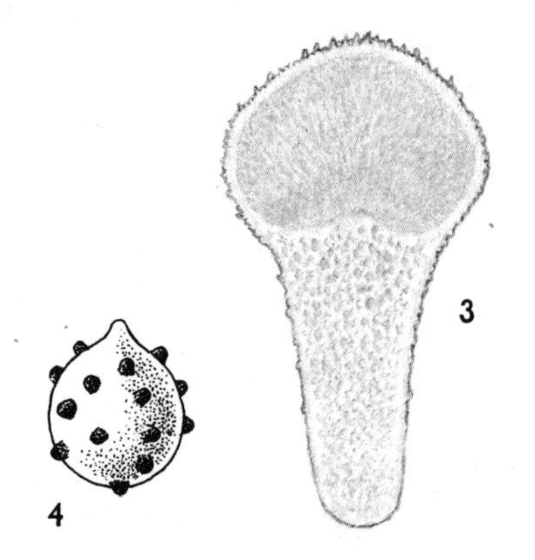


*Lycoperdon perlatum* (1) has fruit-bodies 3—5 cm wide, 2—8 cm high, usually pear-shaped. Exoperidium formed by white or creamy to brown spindle-shaped warts or spines (2). These drop off old specimens and leave a circular pattern in the thin brown-yellow endoperidium (2), which has opening at the apex. Gleba contains brown spore powder and filaments (capillitium). Base is sterile, composed of small chambers (3). Spores (4) yellowish, 2.5—4 μm.

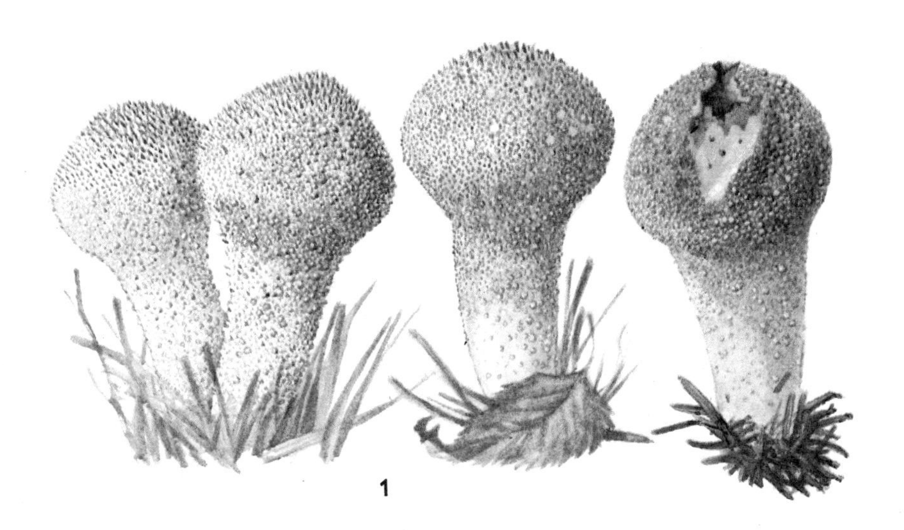
1

*Calvatia utriformis* (5) has fruit-bodies 8—15 cm wide, 5—10 cm high, broadly pear-shaped, flattened at the top; young specimens are white, later grey-ochre. The upper third has regular cracks and later disintegrates into pieces. Gleba contains brown spore powder and filaments. Lower infertile part is bowl-shaped and does not disintegrate (6). Spores brown, 3.5—5 μm.

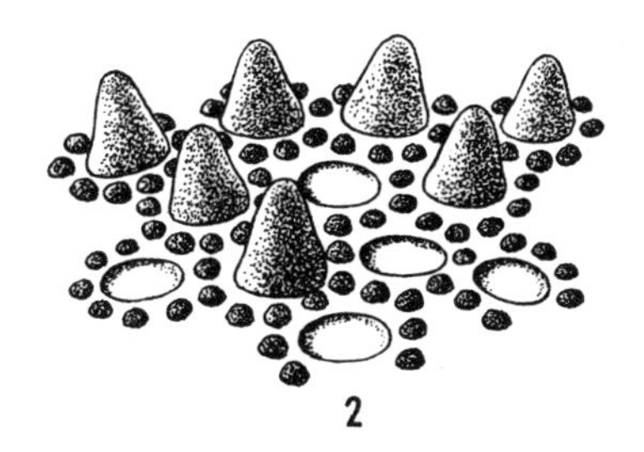
2

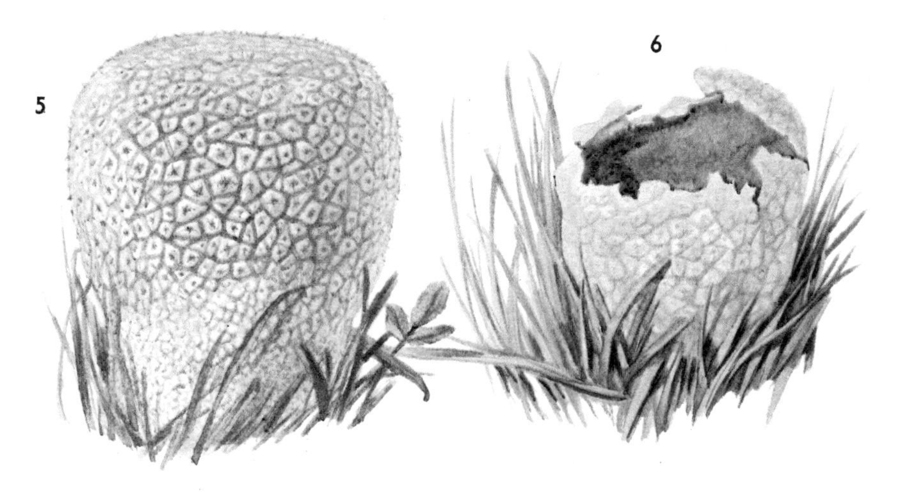
5

6

All members of the genus *Bovista* have globular, sessile fruit-bodies. Young, unripe fruit-bodies are whitish and in this stage they are edible; however, the individual species cannot be distinguished from one another. This can only be done by peeling off the external peridium from mature specimens to reveal the internal paper-like peridium. The endoperidium in *B. plumbea* is grey and matt, while in *B. nigrescens* it is black-brown and glossy. The dry, ripe fruit-bodies can survive in their habitat until the following year. The spores are released through the apical pore. *B. plumbea* and *B. nigrescens* are the most common species of the genus. *B. plumbea* is distributed through the temperate zone of the northern hemisphere and also in New Zealand. In summer and autumn it grows outside forests, in meadows and pastures and along the edges of roads and forests. This species likes a warm climate and is particularly fond of soils with a high content of lime. It grows high up in mountains. The similar *B. graveolens*, a European species, has a grey-brown endoperidium and 10 μm-long stalks (sterigmata) on its spores. These are not straight as in *B. plumbea*, but U-shaped.

*B. nigrescens* grows abundantly in summer and autumn in pastures and meadows and deciduous or mixed forests. It is most prolific in mountains and climbs high up into the alpine zone.

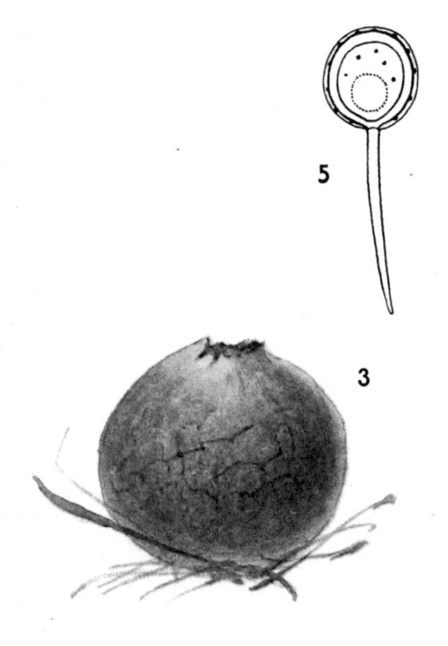


*Bovista plumbea* (1) has fruit-bodies 1—5 cm wide, globular. Exoperidium white, fragile, soon peels away. Beneath is paper-like, blue-grey to grey, matt endoperidium (2) which, when mature, opens through the apical pore (3). Ripe gleba contains chocolate brown spore powder and capillitium (4). Taste and aroma indistinct. Spores (5) brown, 4.4—6.5/4—5.5 μm, with straight sterigmata 8.5—14 μm long. Filaments are olive-brown, thick-walled, without pores (6).

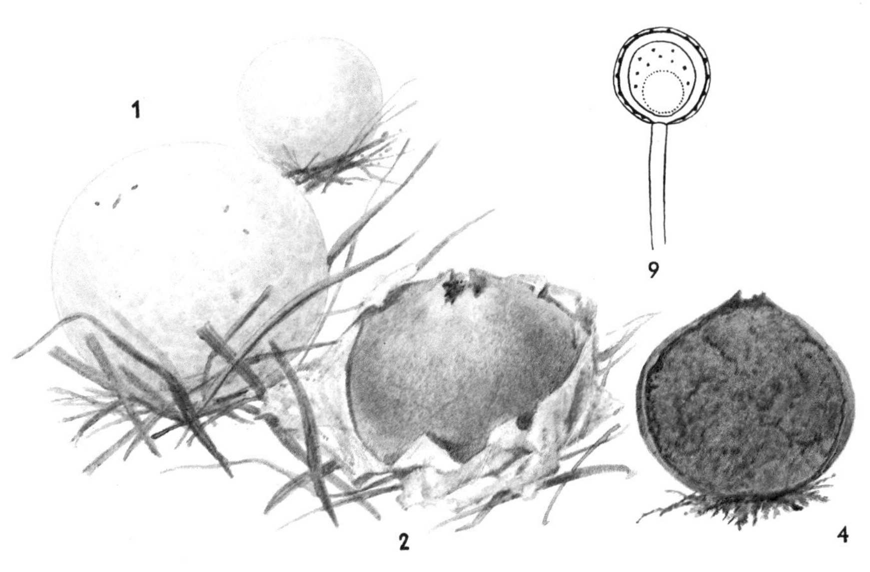


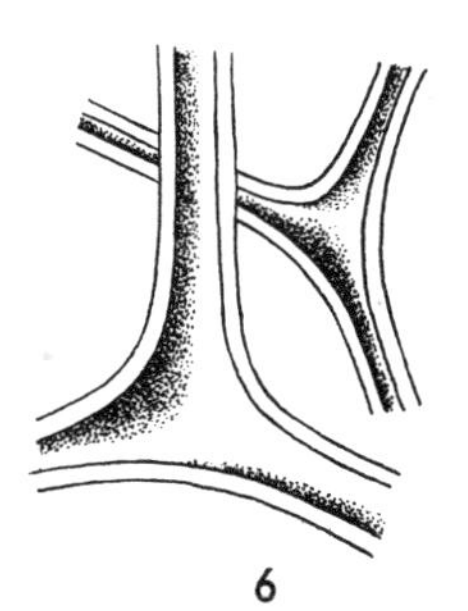


*Bovista nigrescens* (7) has fruit-bodies 3–9 cm wide, globular. Exoperidium white, soon peels off. Endoperidium paper-like, brown to black, glossy, with an apical pore (8). Ripe gleba contains black-brown spore powder and filamentous mass. Taste and aroma indistinct. Spores (9) yellow-brown, 4.5–6 μm, with straight sterigmata 4–9 μm long. Filaments are purple-black, thick-walled; have no pores as in *B. plumbea* (6).

# *Geastrum vulgatum* VITT.
## Syn.: *Geastrum rufescens*



In their appearance, unusual development and botanical character, the earth-stars are among the most interesting of mushrooms. They grow as saprophytes on bare ground or on tree stumps and trunks in an advanced state of decay. They originate as an underground globular formation with surface mycelium. The exoperidium of the ripe sphere bursts into several fleshy lobes, which, when they straighten out, bear the internal fertile layer (gleba) above the ground. The thin, paper-like endoperidium lies on these star-shaped lobes. The circular pore projects through the apex and ejects the spores into the air. The shape of the area surrounding the pore varies according to the species and helps distinguish it. New earth-stars appear on the ground in summer and autumn and last for almost a year because of the toughness of their leathery fruit-bodies. In their young, globular stage the earth-stars are edible but difficult to digest because of their leathery texture.

*G. vulgatum* grows individually in coniferous and deciduous forests throughout the temperate zone of the northern hemisphere. It can be easily recognized because it turns pink.

*G. sessile* is found almost worldwide; it is particularly abundant in coniferous forests. It can be confused with *G. saccatum,* a similar species which is rare but very common in North America. It can be distinguished by its filamentous pore, which is encircled by a border and by its smooth, ochre-yellow exoperidium.

*Geastrum vulgatum* (1,2) has fruit-bodies 2—5 cm high and 3—11 cm wide, consisting of a globular gleba. It contains light brown spore powder and capillitium. Brown exoperidium is sessile and lies on 5—8 fleshy, star-shaped, thick lobes up to 7 cm long, pink in early stages, later brown. Taste and smell are inconspicuous. Spores (3) pale brown, 3.5—4.5 μm.

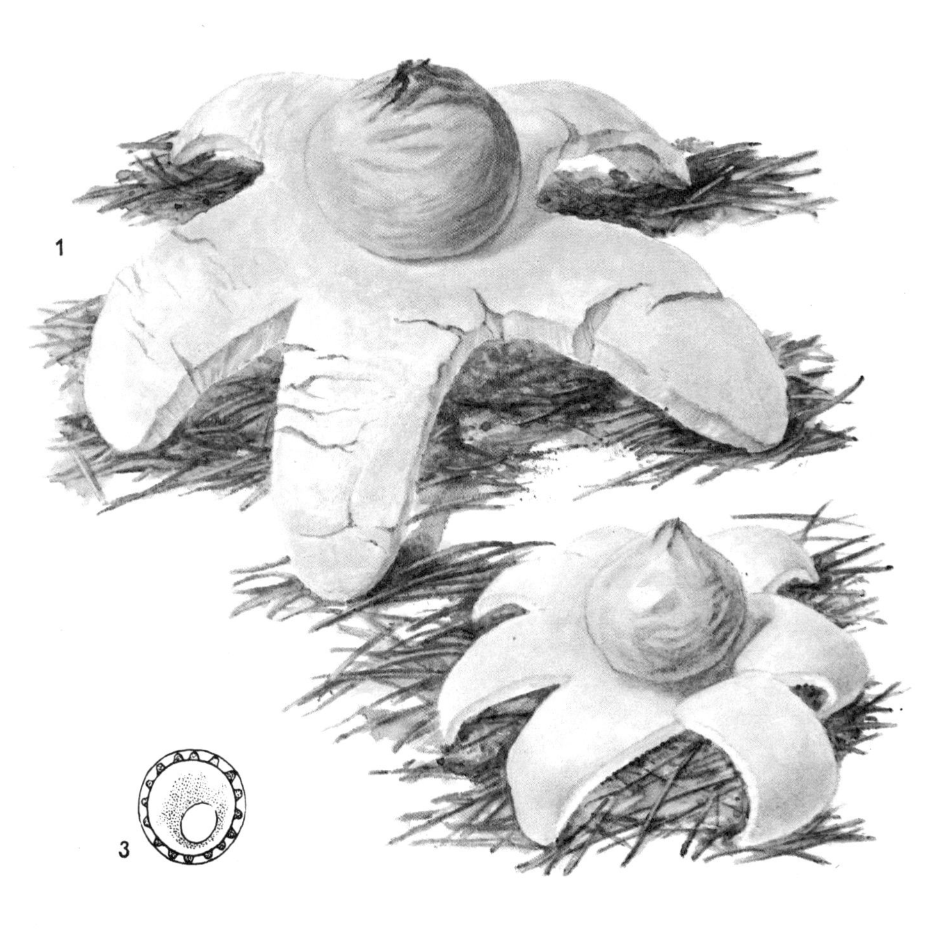

5

Geastrum sessile (4) has fruit-body 1.5—3 cm tall, 2—7 cm wide, consisting of gleba. It contains brown spore powder and capillitium. Brown exoperidium is sessile, lies on 5—10 brownish, downward curved, star-shaped lobes; its lower surface carries mycelium and remnants of humus. Taste and smell indistinct. Spores (5) light brown, 2.5—4 µm.

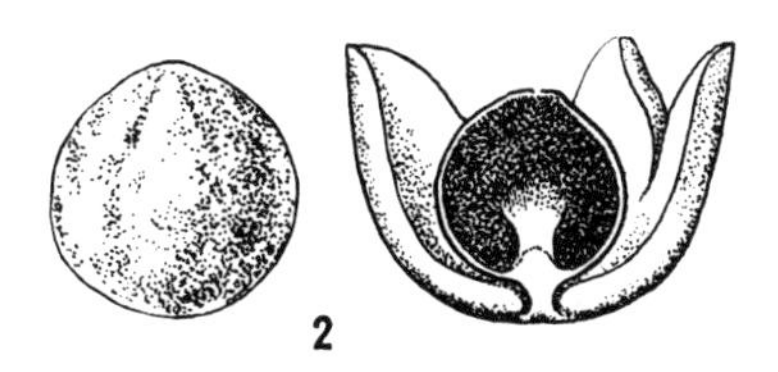

## Ecology of Fungi

Fungi can be found everywhere and in a great variety of habitats. They do not grow in isolation but live with other organisms in natural interdependent communities, in which they have an important and irreplaceable function. It is essential to realize that the characteristic mushrooms, the fruit-bodies which grow in soil, on wood etc, are only reproductive organs, similar to the blossoms of green plants. It is the mycelium which is concealed in the soil or wood that is the most important part of a mushroom. The mycelium of all fungi is similar in appearance and it is still impossible to identify the species from it. Under favourable conditions the mycelium forms fruit-bodies for a period. Such periods are called the fructification phase of the species. Some species fructify only in spring, such as *Hygrophorus marzuolus, Calocybe gambosa* and *Entoloma vernum.* Others form fruit-bodies only in winter, for example *Pleurotus ostreatus* and *Flammulina velutipes,* while others, such as *Armillariella mellea, Lactarius deterrimus* and most *Clitocybe* and *Cortinarius* species grow only in autumn. If the conditions are very advantageous, a mass growth of fruit-bodies takes place, a phenomenon which is called a fungoid aspect. For example, aspects of species of the genera *Russula* and *Cortinarius* are frequent in autumn. Many species fructify from the middle of June until the end of October. During this phase many species have two stages of growth. The first begins in the second half of June and ends at the beginning of July and the second begins in September and lasts until the end of October. In certain years both growing seasons can move forward or backward, grow longer or become shorter, depending on the particular weather. In northern Europe the beginning and end of fructification differ considerably from southern Europe. Some species, however, produce fruit-bodies all the year round and it is quite common for autumn mushrooms also to be found in spring. If, in spring, a warm period is followed by a lengthy period of cool, rainy weather, this climatic change can lead to the appearance of such typically autumnal species as *Armillariella mellea, Lepista nuda* or *Lepista nebularis.* The most important climatic influence in the growth of mushrooms is humidity. The decisive factor is not the macroclimate i. e. the overall weather conditions but the specific microclimatic conditions of each habitat. Humidity is influenced by a number of secondary factors, including geographical location, the shape of the terrain, height above sea level, physical relief and the composition of the soil. Individual species vary considerably in their soil moisture requirements. Aquatic and marsh fungi grow on substrates which are completely submersed in water; these include *Cudoniella clavus* and *Mitrula paludosa;* the peat-bog species are, e. g., *Dermocybe uliginosa, Lyophyllum palustre* and *Galerina paludosa.* Mushrooms which expect an average soil humidity include some of the *Tricholoma, Boletus* and *Agaricus* species. Most wood species require less in terms of substrate humidity. Xerophilous fungi which demand drier soils belong largely to the *Gasteromycetes* group.

The temperature requirements of different species vary similarly. Many are mezzothermic — that is, they grow almost every year in average temperature and humidity. Some species like warm climates and extreme temperatures (xerophilous fungi). These include species of the genera *Geastrum, Bovista* and *Calvatia* and *Boletus satanas* and *Amanita vittadinii.* These species are widespread in southern Europe. Other species prefer a cold climate and therefore grow in northern regions. Among these are *Calvatia cretacea, Omphalina kuehneri* and *Arrhenia auriscalpium.*

Light is much less important for mushrooms than for green plants. It does, however, influence their colouring. Mushrooms which grow in complete darkness are white and without pigment and acquire their typical colouring only when exposed to light. The fruit-bodies of wood species which have developed in the total darkness of mines or cellars are unnaturally branched, extended and white. Some common fungi of the family Polyporaceae were formerly considered to belong to a separate genus when found growing in darkness.

Air currents and wind do not promote fructification; on the contrary, they slow it down because they dry out and cool the soil, and decrease the humidity of the air.

Most fungus species have almost constant requirements as to climate, soil, composition of green vegetation surrounding them, height above sea level etc. (If any of these conditions are unsuitable a particular species will either not grow in such a habitat at all or, if it does grow, will only do so for a short period.) This leads to the existence of mushrooms peculiar to steppes, woodland, meadow, peat-bog, sandy and other local ecotypes, such as spruce stands, oak and beech forests, and mountainous, alpine or arctic environments. Some species require a lime-rich soil, for example *Boletus satanas* and *Entoloma sinuatum;* others prefer an acid soil, such as *Hygrophorus marzuolus* and *Amanita porphyria,* while a number of other species require a neutral soil. Nitrophilous species grow only in soils with a high nitrogen content. These include for instance *Agaricus* species and *Marasmius oreades.* Animal excrements provide a habitat for coprophilous fungi, such as some species of the genera *Stropharia* and

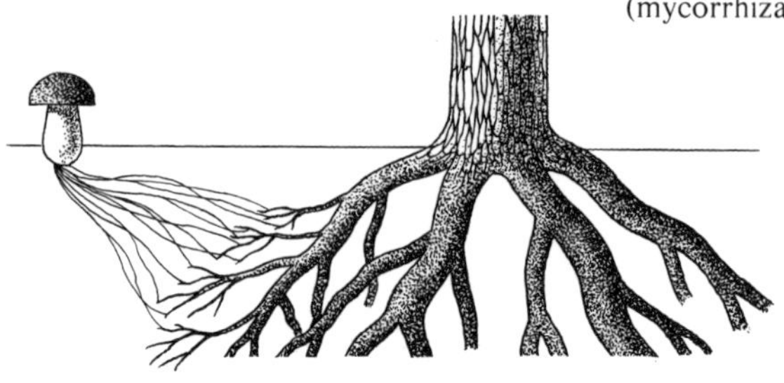

7. Symbiosis between trees and fungi (mycorrhiza)

*Coprinus.* Some species inhabit a wide range of substrates, such as *Mycena pura* and *Collybia dryophila.* However, they are relatively rare. Certain wood fungi, on the other hand, can be strictly dependent on a single tree species. For example *Piptoporus betulinus* grows only in birch trees. Other species, however, are not so specialized, and can be found on a variety of tree species, as is the case with *Mycena galericulata,* for instance.

The substrate which supplies mushrooms with their food is the most important determining ecological factor. According to their way of taking nourishment, mushrooms are divided into parasitic, saprophytic and myccorhizal species. Parasitic species are least common among larger fungi, saprophytic mushrooms, such as the genera *Clitocybe, Mycena* or *Lepiota,* which live on the decaying matter of dead organisms, are the most numerous. The most interesting group, however, are the mycorrhizal mushrooms; these include the *Amanita, Cortinarius, Russula, Lactarius* genera and most *Boletus* species. In ectomycorrhiza the mycelium envelopes the fine roots of trees and partly penetrates them without passing through their cell structure. Fungal hyphae take from the tree a proportion of the nutrients it assimilates and, in return, supply it with or enable it to absorb other nutrients. Their relationship is symbiotic, that is beneficial to both partners. Some mushrooms form ectomycorrhiza with various trees, while others are dependent on particular tree species. *Suillus grevillei* and *Hygroporus lucorum* thus form mycorrhiza only with larch; *Suillus granulatus* does so only with those pine species which have their needles in pairs. This symbiotic association of a mushroom with a higher autotrophic plant is called ectotrophe. It is, in fact, a complex process similar to that found in lichens, where an alga grows most frequently with an *Ascomycete* fungus. It is well known that most mycorrhizal mushrooms follow their host tree everywhere — from lowlands to high mountains, from the temperate belt to the far arctic regions. The existence of ectotrophes in harsh mountainous or arctic conditions is helped by this close relationship based on mutual nourishment. In Europe all forests are ectotrophic, but in the virgin rain forests of Africa and South America ectotrophes are almost completely unknown. The dependence of the fungoid component on the ectotrophe varies. It can be very close, in which case the area of distribution of fungoid and tree component is identical. In other instances the relationship is much looser and the areas of distribution are not necessarily identical. Such information is important when new tree plantations are being established in areas where the wood species forming these ectotrophes are not indigenous. If mushrooms which constitute the other part of the ectotrophe are also missing the plantation will fail. In southern Chile, new plantations of pine trees can be only successfully established when the soil is impregnated with the mycelium of the Yellow-brown Boletus mycelium, while in Argentina the Granulated Boletus is used for the same purpose. Countries which export trees to foreign countries often supply the seedlings together with the mycorrhizal mushrooms.

The distribution of mushrooms on the Earth's surface is very uneven. The richest mycoflora can be found in the temperate zone of the northern hemisphere, most of all on the North American continent, followed by Europe. Research into the mycoflora of the tropical zones of the old and new worlds is still incomplete, but this mycoflora is likely to be very rich indeed. As a result of the ease with which their spores are propagated by air currents, fungi have a much wider area of distribution then green plants. The mycoflora of all European countries is therefore very similar and, at the same time, there are many species common to both the North American and European continents. The majority of the European species illustrated here grow both in North America and north Africa. However, the tropical mycoflora of these continents does differ from the species of Europe.

## Identifying Fungi

The identification of fungi is rather difficult, because a knowledge of common morphological features must be supplemented by a knowledge of anatomical, chemical and other characteristics. While the morphology of mushrooms can be quite adequately studied with the naked eye or with the help of a magnifying glass, anatomical and spore analysis definitely requires a microscope. Basic observation requires simple microscopes with an overall magnification of 600. Particular microscopic preparations such as spores, hyphae and cystidia can be observed either in a water solution, or if some specimens require sharper delineation, a solution of aniline (cotton) blue in lactic acid or Melzer's reagent may be used.

The simplest chemical reaction occurs when a fruit-body is cut in half and the oxygen in the air oxidizes its tissue and changes its colour. The flesh either turns blue, red, yellow or black. Similar changes occur when the surface of the fruit-body is scratched or when the tubes or gills are bruised. The colour changes in the milk of *Lactarius* species are also typical results of oxidization. The tissue of some fungi, particularly *Russula* species, changes colour when it comes into contact with certain chemicals. The most frequently employed is a solution of green vitriol ferrous sulphate and a 5% solution of potassium hydroxide. For certain identification, it is important to describe correctly not only the colours of all parts of the given mushroom, but also its taste and smell. Tasting is carried out using the tip of the tongue only and the sample should never be swallowed.

Only a responsible and serious approach to the identification of an unfamiliar mushroom can safeguard against confusion with a poisonous species.

## Mushroom Picking

Mushrooms are collected mostly for cooking but also for scientific research. They can be gathered all year round, even in winter. It is always advisable to consider the environment. Moss or humus should never be dug away, as this causes the soil to dry out quickly and the mycelium of mushrooms to die. Unknown species should never be destroyed. They have their own ecological function, as well as serving as a natural decoration.

Mushrooms should be placed in well-ventilated containers, baskets or boxes, but never in plastic bags, where they perspire. The fruit-bodies should be carefully removed with a slight twist and the disturbed ground covered with humus. Mushrooms should be carefully examined. If recognized beyond doubt, they may be cleaned and placed alongside other edible species. If there is any doubt a mushroom should not be cleaned, but wrapped undamaged and stored away from edible species. It can then be identified at home, using a guide or key. If initial identification has been made without the use of a microscope, the findings should be confirmed by an expert. Old, soft or woody fruit-bodies or mushrooms attacked by insect larvae should not be collected.

When collecting mushrooms it is important to wear suitable clothing for the weather and the time of year. Solid shoes, long trousers and a shirt with long sleeves even in summer can provide protection against parasites and dangerous insects. A good quality folding knife is indispensable.

It is best to go mushroom picking early in the morning. There is more chance of a good collection at this time of day, as the mushrooms should be free of maggots and will be fresh after the cool of the night. They will therefore more easily survive being transported. Firmer mushrooms should be placed at the bottom of the basket and more fragile ones at the top. The ground should be studied systematically and with steady concentration so as not to miss mushrooms which are usually hidden and well camouflaged by their surroundings, with which the colour of the cap often blends. If a mushroom is found, its surroundings should not be disturbed in a further attempt to find other specimens; this damages both the humus and mycelium and limits further growth.

It is impossible to give precise information about the times when different species of mushroom grow. The basic requirements for growth are moisture, warmth and a still atmosphere. Most mushrooms only grow several days after heavy rainfall, and only if this is followed by a warm spell. A popular saying among mushroom pickers, 'when forests steam, mushrooms thrive', has some scientific basis. According to the season of the year and the type of forest the possibility of finding certain species varies. The secret of successful mushroom picking is a long-standing knowledge of forests and other locations. It is better to know a small number of species well, rather than know a large number superficially. Only species which are recognized beyond any shadow of doubt under all conditions should be collected for eating. If at all in doubt an expert should be consulted before the mushroom is eaten. Poisoning caused by mushrooms is very unpleasant and can result in death. When treating a case of mushroom poisoning it is essential to eliminate the offending poison from the body as quickly as possible. This can be done before the doctor arrives by administering an emetic.

Mushrooms should be picked in quantities which can be immediately used. Fresh mushrooms are best cooked the same day. Failing this they can be stored in a refrigerator at a temperature of about 5°C for a maximum of three days. Mushroom dishes decay very quickly but they too can be stored in a refrigerator and reheated the following day without danger.

Collecting mushrooms for scientific purposes is more demanding and the processing of specimens takes a long time. Specimens should be whole, fresh and undamaged. Fruit-bodies from the same mycelium should be gathered and stored separately, and the collection should contain samples of young, adult and old fruit-bodies. Each specimen should be wrapped, separately, or in some cases several together, in a newspaper or soft tissue paper or aluminium foil and then placed in a firm box. Directly on the spot, notes should be made, describing the general characteristics of the habitat, associated green plants and trees and the substrate in which the mushrooms were found. If the mushrooms cannot be quickly identified, precise notes are needed, as even the most careful storage damages many important features. The notes should describe the overall appearance of the mushroom, as well as its colour, taste and smell, and any colour changes noted in its cross-section. Sketches should accompany the written description. The most common conservation method is by drying with warm air at a temperature of between 40° and 50°C. The fleshy fruit-bodies are cut into sections. The dry fruit-bodies, known as exsiccates, are then stored in envelopes or small boxes, and a mushroom herbarium is built up. All the dried specimens must be thoroughly disinfected, for example with carbon disulphide, otherwise they may be attacked by insects. Another method of conserving mushrooms is by using various liquids. For example, fresh mushrooms can be submersed in a mixture of water, ethanol and formaldehyde. This maintains their shape and to some degree their colour.

## Cultivation and Nutritional Value of Mushrooms

Mushroom picking in the wild is a popular pastime in many countries. However it accounts for only a small proportion of the consumption of edible mushrooms. Gathering edible mushrooms in municipal parks or near large industrial estates is not recommended, as they will often contain a high concentration of poisonous substances absorbed from the atmosphere or from chemically-treated trees and soil. Mushrooms growing in forests alongside highways may be harmful for similar reasons. For these reasons countries regulate the commercial collection of wild mushrooms strictly. The artificial cultivation of many edible species has therefore become very important and the production rate in many countries rises every year. For example in 1950 the world production of cultivated Field Mushrooms was about half a million tonnes; current production is estimated at almost one million tonnes.

Opinions are varied as to the nutritional value of mushrooms. Some scientists have overestimated their importance as a human food source, equating them with meat and eggs; other specialists have dismissed them as nutritionally worthless, and completely indigestible because of their high content of mycochitin (a substance which forms the cell lining and is chemically related to the chitin of arthropods). It is well-known that mushrooms have a very low caloric value. However, they create a feeling of satiation even when eaten in small quantities and this is an important factor in weight-reducing diets. In salads, sauces, soups and aromatic additions to meat dishes, they form an important supplement to any diabetic menu. As opposed to meat, which contains 50—70% water, mushrooms contain on average about 90% water. They contain very little sugar, 1 to 1.5%, but some valuable protein, 4—5%, of which 70—80%, can be digested. Mushrooms can even surpass meat in taste and aroma and they also favourably compare with vegetables and fruit. They contain a number of vitamins and minerals; namely provitamin A, a group of B vitamins, vitamin D and a very small amount of vitamin C. Mineral substances, which constitute 1—1.5%, include potassium, calcium, magnesium, iron, phosphorus, traces of fluorine, copper, manganese, cobalt and titanium. Various aromatic substances are a valuable constituent of mushrooms and are found in significant quantities. Mushrooms also contain about 0.5% of fats, as opposed to 5—10% in meat. This can be observed in the form of droplets of oil on the spores. Like cellulose in vegetables, the indigestible mycochitin element in mushrooms has a beneficial effect on intestinal peristalsis and therefore the digestive process. From the nutritional point of view, therefore, mushrooms must be favourably regarded. They are not simply a delicacy or spice, but a natural food with valuable nutritive qualities. Thus the rapid increase in mushroom cultivation in recent years is therefore a positive development. The first attempts to cultivate mushrooms were made in Japan and China in the late second and early third centuries AD. These experiments involved the wood mushroom *Lentinus edodes,* known in Japan as 'shiitake', which grows naturally in the temperate and subtropical zones of eastern Asia. Today the same mushroom is cultivated in Japan in large quantities, with an annual production of approximately 120 thousand tonnes. *L. edodes* also has medicinal properties as it reduces the level of cholesterol in the blood and may slow down the hardening of the arteries. In the tropical regions of south-eastern Asia and on Madagascar, *Volvariella volvacea,* which also grows in Europe, is traditionally cultivated on beds made of rice straw. In recent years this mushroom has also been grown in Japan. In China and Japan 10 thousand tonnes of the Jew's Ear Fungus *(Hirneola auricula-judae)* are produced annually, and other cultivated species include *Pholiota nameko, Tricholoma matsutake, Tremella fuciformis* and some *Auricularia* species. The Oyster Fungus *(Pleurotus ostreatus),* the Winter Fungus *(Flammulina velutipes),* Changing Pholiota *(Kuehneromyces mutabilis)* and *Agrocybe aegerita* have been successfully cultivated in many countries. These mushrooms can be grown by amateurs; the wood host, when colonized by the mycelium, needs periodical watering, and fructification occurs when favourable conditions are reached. Another species suitable for amateur cultivation is *Stropharia rugosoannulata,* which can be grown on straw in the garden and in a hot frame. Experiments are continually being made with other species, including *Coprinus comatus* and *Polyporus tuberaster.*

The leading species cultivated worldwide are the field mushrooms *Agaricus bisporus* and *A. bitorquis.* The first large-scale cultivation of this mushroom was in France in the middle of the seventeenth century. In 1707 Count Tournefort published the first description of field mushrooms in the proceedings of the French Academy of Science. France kept the lead in the cultivation of mushrooms for a long time but was overtaken by the United States of America during the World War II.

Field mushrooms are not only grown in disused underground sites, but also in special above-ground nurseries. The success of modern methods of bulk production of edible mushrooms depends primarily on the high yield of mushroom cultures. This means in effect that a high protein crop is obtained without any demands on farmland. The annual yield of field mushrooms from one square metre is between 50 to 120 kg and if they are grown on five levels or more, the same meterage will produce up to 600 kg. By way of comparison the same area of soil will yield 3—5 kg of potatoes, 0.3—0.5 kg of wheat or 15—35 kg of cucumbers. The prime producer of field mushrooms is now the United States of America with a yearly turnover of about 120 thousand tonnes; followed by France with 100 thousand tonnes and Taiwan with 85 thousand tonnes.

All bulk-grown mushroom species are saprophytic and therefore do not need the presence of trees or other plants. Boleti and other mycorrhizal mushrooms have not been cultivated successfully because they require the presence of their symbiotic partner. However some species of mycorrhizal mushrooms have been grown by impregnating the roots of the appropriate tree with the mycelium of the relevant mushroom. The mycelium then develops with the tree and produces fruit-bodies if given suitable environmental conditions. In this way it has been possible to grow some *Suillus* species under pine trees. In France this principle has been employed in constructing truffle nurseries. Young oak seedlings are impregnated with laboratory cultures of *Tuber melanosporum.* After six years the well-flavoured, sharply aromatic truffle fruit-bodies start developing in close proximity to the roots of the young oak trees. Trained pigs or dogs are then used to locate these delectable fungi. However, this method cannot be considered true mushroom farming but simply reinforcing natural growth.

## Mushroom Poisoning

For centuries mushrooms have been picked for their excellent flavour. But as well as a large number of edible mushrooms there is a high percentage of poisonous species, causing hundreds of cases of poisoning every year, some ending fatally. There are about a hundred poisonous species in Europe, out of which eight are deadly poisonous. Probably the most poisonous mushroom is *Galerina sulciceps,* a wood species which grows in Java and Ceylon. After eating a single mushroom the victim dies within 7 to 51 hours. The most poisonous mushrooms in Europe and North America are considered to be the Death Cap, Spring Amanita and Destroying Angel. Poisoning was followed by death in 90% of cases, but today the mortality rate is about 40%. Toxic agents develop in mushrooms as a specific by-product of their metabolism. These substances can be chemically analysed and used in an experimental reconstruction of the poisoning process as well as in experimental treatment.

Mushrooms which are edible in their early stages can become toxic with ageing, due to the presence of poisonous microorganisms, or as a result of being sprayed with pesticides or herbicides. If the mushrooms grow along the roadside, they can absorb poisonous heavy metals such as mercury, lead or cadmium. Symptoms of mild poisoning can also occur together with certain infections, temporary indispositions or psychological reactions after the consumption of a large amount of a particular mushroom dish. Some mushroom species which are edible after being cooked contain poisonous substances in their raw state. These include the Honey Fungus, Lurid Boletus and others. Only a few mushrooms can be safely eaten when raw, such as *Lactarius volemus,* the Jew's Ear Fungus, the Edible Boletus, *Boletus aestivalis, Boletus pinicola* and *Pseudohvdnum gelatinosum.*

Poisoning takes a more serious course and results in a much higher death rate with children. Children should avoid eating raw mushrooms, large quantities of cooked mushrooms or dishes prepared from mushrooms of a poor quality. However, there is no general rule to distinguish poisonous mushrooms from edible ones. The only safe defence is a sure knowledge of the mycological features of edible and poisonous mushrooms and an awareness of their differences. The basic principle should be to collect only those species which are well known and which can be reliably identified in all conditions, whether they are young or old, wet or dry.

Types of poisoning can be determined according to the group of mushrooms involved and their poisonous attributes:

1. Mushrooms containing deadly poisonous cyclopeptides (phallotoxins) include various *Amanita* species *(Amanita phalloides, A. verna, A. virosa, A. bisporigera), Galerina marginata* and *G.autumnalis* and some small *Lepiota* species, especially *L. helveola.*

The first symptoms occur between 6 to 24 hours, or sometimes even after 48 hours. They are characterized by acute diarrhoea, vomiting, urination, spasms and thirst. About three days after consumption there is a period of apparent relief, but this is soon followed by jaundice and the patient dies through malfunction of the liver. Medical help should be sought immediately.

Poisonous mushrooms of this group are often mistaken for edible species. For example the Death Cap is confused with the green *Russula* species, *Tricholoma flavovirens* and *T. portentosum.* The Spring Amanita and the Destroying Angel can be taken for field mushrooms, *Galerina marginata* for *Kuehneromyces mutabilis, Armilariella mellea* or even *Flammulina velutipes.*

2. Mushrooms containing the haemolytic poison monomethylhydrazine include some Brain-fungi *(Gyromitra esculenta* and probably *G. fastigiata).*

The first symptoms occur 6 to 12 hours after consumption, sometimes as early as 2 hours. Tiredness, headache, dizziness, sickness, and vomiting lasting 1 to 2 days are followed by jaundice and liver discomfort, sometimes with fatal results. The course of this poisoning is very similar to the one described in the first group. Treatment consists of B group vitamins and muscle relaxants.

Poisonous mushrooms of this group are often mistaken for the edible *Morchella esculenta.* Since this poison is thermolabile, the mushrooms lose their toxicity if they are boiled for 15 minutes and then strained.

3. Mushrooms containing the poisonous substances orellanin,

grzymalin and cortinarin belong to the genus *Cortinarius (Cortinarius orellanus, C. speciosissimus* and related species*).*

The first symptoms of poisoning become evident only after 2 to 14 days and sometimes even later. These include increased urination, a dry feeling in the mouth, stomachache and vomiting. Kidney malfunction follows and the patient dies. There is no specific treatment for this poisoning, but it is important to maintain kidney activity.

Poisonous mushrooms in this group are mistaken for the various edible *Cortinarius* species.

4. Mushrooms containing coprin include various *Coprinus* and *Clitocybe* species, the Common Ink Cap *(Coprinus atramentarius),* probably the Glistening Coprinus *(C. micaceus)* and the Club-foot Clitocybe *(Clitocybe clavipes).*

If alcohol is consumed, even two days after eating these mushrooms, symptoms of poisoning appear within 30 minutes. The poisoning is characterized by reddening of the face and body, a strong heart beat, stomachache, vomiting and diarrhoea. These symptoms usually last between 2 to 4 hour and, after further consumption of alcohol, the same reaction can be repeated several times. This poisoning is not fatal.

Young fruit-bodies of *Coprinus* species are eagerly sought by connoisseurs, because of their delicious flavour. Interestingly, a similar poisoning process described above has been used for the treatment of alcoholism.

5. Mushrooms containing the poisonous alkaloid muscarine belong to several genera. The first in importance is the *Amanita* genus, followed by *Inocybe patouillardii* and many more of the *Inocybe* genus. The white *Clitocybe* species *(Clitocybe dealbata, C. cerussata),* the *Mycena* genus *(Mycena rosea)* and the *Omphalotus* genus *(Omphalotus olerius)* are also included. Muscarine was first found in *Amanita muscaria* (hence the name), but the species contains so small an amount of it that it doesn't cause poisoning. The first symptoms occur between 30 minutes and 2 hours after consumption. The face turns red and the victim perspires and salivates. This is followed by a fever without rise in temperature, respiratory disorders, defects of vision and stronger heart beat. The poisoning can be successfully treated with atropine.

Poisonous mushrooms in this group can also be confused with some edible species. For example *Inocybe patouillardii* can be mistaken for *Calocybe gambosa,* small *Inocybe* and white *Clitocybe* species with *Marasmius oreades, Mycena rosea* with *Laccaria* and *Omphalotus olearius* with *Cantharellus cibarius.*

6. Mushrooms containing such toxins as iboten acid and muscimol (both formerly thought of as mycoatropin) include various *Amanita* species *(Amanita muscaria, A. regalis, A. gemmata* and *A. patherina).*

The first symptoms occur between 30 minutes and 2 hours after consumption. The heart beat increases and there is mild perspiration, followed by restlessness and typical symptoms of alcoholic intoxication. These symptoms cease after 1 to 2 hours and there is no treat of death. Hallucinations can occur after consuming certain geographical varieties of the Fly Agaric.

Poisonous mushrooms of this group can be mistaken for *Amanita rubescens* and *A. spissa.*

7. Hallucinogenic mushrooms containing psilocybin and psilocyn include *Psilocybe semilanceata* and many other species belonging to the *Psilocybe* genus which mostly grow outside Europe. There is also *Panaeolus fimicola, P. feonisecii, P. sphinctrinus* and probably also *Mycena pura* and *Amanita muscaria* (the latter two containing indole substances).

The first symptoms occur between 30 and 60 minutes. After ingestion pleasant visual and aural hallucinations last for 1 to 2 hours, but there is no danger of death.

Hallucinogenic mushrooms were formerly eaten by Mexican and South American Indians during their tribal rituals. They were collectively known under the name of 'teonanacatl'. Psilocybin is used in modern medicine for psychiatric disorders.

8. A mushroom which causes allergic poisoning in sensitive individuals is the Inrolled Paxil *(Paxillus involutus).* Symptoms of this poisoning can occur as soon as several hours or as long as several years after consumption, depending on the sensitivity of the individual. Dizziness, stomachache, diarrhoea and chest pains are followed by the appearance of blood in the urine. Subsequent malfunction of the kidneys can lead to death. Treatment is directed towards maintaining kidney function.

The Inrolled Paxil used to be considered an edible mushroom and was collected in large quantities and is therefore included as an edible mushroom in old mushroom guides.

9. Many fungus species cause gastrointestinal disordes. These include some species of the genus *Agaricus (Agaricus xanthodermus* and related species), some *Boletus* species in their raw state *(Boletus luridus* and *B. satanas), Chlorophyllum molybdites, Entoloma sinuatum, E. vernum* and also *Lactarius helvus, L. torminosus* and *Hypholoma fasciculare.*

The first symptoms of poisoning occur between 30 minutes and 2 hours after consumption. They take the form of sickness, headache, stomach pains, dizziness, vomiting and diarrhoea. But such poisoning is rarely fatal.

Poisonous mushrooms of this group can be confused with related edible species.

10. Mushrooms containing poisonous alcaloids of the type of bufotenin are certain *Amanita* species *(Amanita citrina* and *A. porphyria).* Poisoning occurs only after eating large quantities of these mushrooms.

## Index